Inventive Strategies for Teaching Mathematics

Implementing Standards for Reform

PSYCHOLOGY IN THE CLASSROOM: A SERIES ON APPLIED EDUCATIONAL PSYCHOLOGY

A collaborative project of APA Division 15 (Educational Psychology) and APA Books.

Barbara L. McCombs and Sharon McNeely, Series Editors

Series Titles

Becoming Reflective Students and Teachers With Portfolios and Authentic Assessment—Paris & Ayres
Creating Responsible Learners: The Role of a Positive Classroom Environment—Ridley & Walthers
Inventive Strategies for Teaching Mathematics—Middleton & Goepfert
Motivating Hard to Reach Students—McCombs & Pope
New Approaches to Literacy: Helping Students Develop Reading and Writing Skills—Marzano & Paynter
Overcoming Student Failure: Changing Motives and Incentives for Learning—Covington & Teel

In Preparation

Teaching for Thinking—Sternberg & Spear-Swerling
Designing Integrated Curricula—Jones, Rasmussen, & Lindberg
Effective Learning and Study Strategies—Weinstein & Hume
Positive Affective Climates—Mills & Timm
Dealing With Anxiety in the School—Tobias & Tobias

Inventive Strategies for Teaching Mathematics

Implementing Standards for Reform

James A. Middleton and Polly Goepfert

AMERICAN PSYCHOLOGICAL ASSOCIATION | WASHINGTON, DC

Published by
American Psychological Association
750 First Street, NE
Washington, DC 20002

Copies may be ordered from
APA Order Department
P.O. Box 2710
Hyattsville, MD 20784

In the UK and Europe, copies may be ordered from
American Psychological Association
3 Henrietta Street
Covent Garden, London
WC2E 8LU England

Typeset in Berkeley and Bell Gothic by University Graphics, Inc., York, PA
Printer: Data Reproductions Corporation, Rochester Hills, MI
Cover Designer: KINETIK Communication Graphics, Inc., Washington, DC
Technical/Production Editor: Edward B. Meidenbauer

Library of Congress Cataloging-in-Publication Data

Middleton, James A.
Inventive strategies for teaching mathematics : implementing standards for reform / James A. Middleton and Polly Goepfert.
p. cm. — (Psychology in the classroom)
Includes bibliographical references.
ISBN 1-55798-368-2 (alk. paper)
1. Mathematics — Study and teaching. I. Goepfert, Polly. II. Title. III. Series.
QA11.M4964 1996
510′.71—dc20 96-13468
CIP

British Library Cataloguing-in-Publication Data

A CIP record is available from the British Library.

Printed in the United States of America
First Edition

TABLE OF CONTENTS

vii Preface

1 Introduction

A True Story: Rachel
A Call for Change: NCTM Reform Standards
The *Curriculum and Evaluation Standards for School Mathematics*
The *Professional Standards for Teaching Mathematics*
The *Assessment Standards for School Mathematics*
Statement of Rationale and Goals

17 Goal One: **Building a New Understanding of the Nature of Mathematics**

A Vignette
Social Constructivism: Knowing and Doing

25 Goal Two: **Building and Choosing Curricula**

A *Traditional* Approach to Volume
A *Context* Approach to Volume
The Nature of Mathematical Activity
Principles for Choosing and Designing Curricula
The Role of Models in Teaching Mathematical Concepts
Self-Directed Activities: Thinking About the *Curriculum Standards*

55 Goal Three: **Building Teaching Strategies**

Thinking About Teaching Philosophies
Planning
Questioning
Self-Directed Activities: Thinking About the *Teaching Standards*

73 Goal Four: Building a Balanced Assessment Strategy

A Personal Note from Polly: Choosing a Strategy
Assessment versus Evaluation
A First-Grade Teacher's Reflections
Assessment Assumptions
Benchmark Opportunities
Assessment Events
What About Grading?
Self-Directed Activities: Thinking About the *Assessment Standards*

113 Goal Five: Finding Resources to Help
Innovative Curriculum Projects
Elementary Projects
Middle-School Projects
High School Projects
Assessment
Technology
Parents
Family Math Nights
Teacher Teams
Administration and the School Board
The National Council of Teachers of Mathematics

147 Reflections
A True Story: Rachel Part II
Conclusion

153 Glossary

157 References

165 About the Authors

PREFACE

> *So, basically, I'm supposed to . . . create a rich and safe environment where all students develop mathematical power by working actively together in heterogeneous groups doing meaningful and worthwhile mathematical tasks using state-of-the-art technology as an appropriate problem-solving tool at the appropriate time, to reach deeper levels of understanding than ever before, largely because of my carefully timed and well-phrased probing questions, while I convince my fellow teachers that this is how they, too, should teach, and while I sell the idea that this is how it really should be to parents, taxpayers, administrators and school board members, most of whom think I should really be spending my time raising my test scores by just doing a better job of what we used to do 20 year ago, in an environment that most of them wouldn't dare set foot into?*

From remarks made by Cathy Seeley at the 1994 National Council of Teachers of Mathematics Annual Meeting. Cathy wishes to express her deep appreciation to the many mathematics teachers who are trying to do just this. By the way, the answer is *yes*.

This book is a reflection of the struggle of two teachers deeply concerned with reforming their own thinking and practice in the field of school mathematics. Change in any domain is difficult, and change in mathematics education is made more difficult with the multiplicity of interested parties making suggestions, recommendations, and mandates that we, as teachers, are expected to implement. This is not to say that recommendations for reform are untenable; quite the contrary. There are fundamental reasons that such recommendations are made. The world is different now than it was when we were schoolchildren. The nature and mission of public schools has broadened. Technology has infused nearly every aspect of our lives.

Our knowledge of how students learn has deepened. The field of mathematics itself is undergoing dramatic change. Traditional shopkeeper arithmetic, appropriate for the low-tech past, has become less and less important in an information age. These and other reasons for rethinking the nature of school mathematics have spurred the reform movement.

The basic philosophy underscoring this book is that implementating reform is a process—that is, standards are visions to work toward, not end points to reach. Moreover, as with any set of standards, recommendations for reform are continually shifting. We learn as we grow, setting ever higher standards for ourselves, our students, and our society.

The purpose of this book is to explore ways in which recommendations for reform can be approached and implemented in real classrooms by knowledgeable and reflective teachers. As you work through this book, (and we mean *work*, not just *read*—this is meant to be interactive, stimulating you to analyze your own experiences, philosophy, and practice) attempt to develop goals for initiating or sustaining your own change process. Find ways to begin change, first in your own classroom and then at the department level, working with other teachers going through the same struggles. With such a collaborative effort, we can effect more global change.

Most important, this book is about *teaching*, that most wonderful and rewarding process by which we engage in meaningful discourse about important and difficult issues that leave a legacy of inquiry and autonomy with our progeny—our students.

James A. Middleton
Polly Goepfert

introduction

This introductory chapter describes the major standards for the reform of mathematics education in the United States. We situate the major documents outlining the vision for change in the historical, technological, and philosophical trends that underline their development. We then outline the rationale and goals of this book in terms of the three major themes of these documents: curriculum, teaching, and assessment. But because we feel that learning should be an emotive, human process, we begin with a true story . . .

A TRUE STORY: RACHEL

From the first author's journal:

> When I first noticed Rachel, a seventh grader, Polly and I had been working together for 2 weeks piloting an innovative curriculum project for all students. But we did not reach all of the students. I had been careful to spend time with each of the cooperative groups in the classroom, gathering information, helping students problem solve, and consulting with the teacher. One day, I swept the room with my eyes, and I spied a little girl alone in the corner, chin in her hands, staring blankly at the front wall. I approached her and posed a question to find out what she was learning. Rachel had not even tried any of the problems that the rest of the class was so excited about. She responded, "I can't do math. I'm bad." Not, "I'm bad at math," just, "I'm bad." Rachel had had a history of child abuse, poverty, physical ailments, and behavior problems. The message she had received up to this point had led her to believe that she was not able to do mathematics. She was isolated, turned off, powerless mathematically, and, as we later discovered, *brilliant*.

Readers should keep Rachel's case in mind as they work through the book; we will return to her at the end.

SELF-DIRECTED QUESTION

1. What would you do to help Rachel if you were her teacher?

A CALL FOR CHANGE: NCTM REFORM STANDARDS

In 1989, the National Council of Teachers of Mathematics (NCTM), the nation's largest organization of mathematics educators, published the first of (currently) three documents outlining a vision for what mathematics education should look like as we progress into the 21st century. These documents, the *Curriculum and Evaluation Standards for School Mathematics*, the *Professional Standards for Teaching Mathematics*, and the *Assessment Standards for School Mathematics*, were designed to reflect the mathematics and pedagogy requisite for success in our increasingly complex and information-oriented society (NCTM, 1989, 1991, 1995).

The main purpose of this book is to help mathematics teachers apply the philosophies behind these standards and implement the suggested reforms in their own classrooms.

Referred to as "*the Standards*" throughout this book, these documents were designed to reflect the important changes that have taken place over the past century that impact on how mathematics is important for professional and economic success and how mathematics should be applied and used by a literate society. Several documents were instrumental in stimulating the mathematics education community to action:

- *An Agenda for Action: Recommendations for School Mathematics of the 1980's* (NCTM, 1980).
- *School Mathematics: Options for the 1990's. Chairman's Report of a Conference* (Romberg, 1984).
- *A Nation at Risk: The Imperative for Educational Reform* (National Commission on Excellence in Education, 1983).
- *The Mathematical Sciences Curriculum K–12: What Is Still Fundamental and What Is Not* (Conference Board of the Mathematical Sciences, 1983a).
- *New Goals for Mathematical Sciences Education* (Conference Board of the Mathematical Sciences, 1983b).

Although prepared by different agencies and written from different perspectives, all of these documents lead toward the same conclusions. The world of our children is very different from the world in which we grew up. The fields of technology have opened new horizons of exploration and inquiry. New mathematical models made possible by technologies are being applied to business, industry, and the professions, and these new mathematics are impacting directly on the lives of the average citizen. Case in point: Since the advent of electronic computing technology in the 1940s, the number of employment opportunities have lessened in the agricultural and industrial fields and have increased in the service and information fields. Moreover, even in the fields of agriculture and industry, robotics, quality control, computerized inventory databases, and other new methods for increasing productivity have changed dramatically the nature of work. New types of knowledge, therefore, are necessary to prepare for successful careers in fields that require flexible use of mathematical knowledge.

The next time you pick up a newspaper, skim through it and take note of the kinds of mathematics you need to make sense of the different articles. You most likely will find lots of *statistics*. The field of statistics is relatively new to mathematics, but since the turn of the 20th century (spurred on by the technological revolution) it has evolved to a point where it may be the most common field of mathematics we as a society use. Opinion polls, scientific studies, lottery commissions, crime laboratories—all of these information sources make use of statistics, and more important, they communicate important data to the public through statistical tables, graphs, and terminology. These and other changes in the field of mathematics demand a new type of knowledge for success, a knowledge of data and patterns, with an eye toward the possibility of multiple interpretations of the same data.

Dramatic changes in learning theory also have broadened our thinking about what mathematics is, what types of mathematical thinking are possible, and as a consequence, how mathematics should be taught. A relatively new philosophy of how humans learn suggests that we do not simply hear and understand. Rather, we attend to certain aspects of a situation based on our prior knowledge and *build* new knowledge structures (we may call these *memories* or *ideas*) that account for the new information. Because knowledge is built, or constructed, by each individual, this approach to learning is called *constructivism* (see Steffe, 1988; von Glasersfeld, 1985). In addition, there is considerable evidence that we learn more, and our knowledge is more useful, when we engage in activities that require us to communicate with others. (For a discussion of the costs, benefits, and possible directions for cooperative group work in mathematics, please see Good, Mulryan, and McCaslin, 1992). Argument, debate, and mutual struggle through difficult ideas allows the strengths of different individuals to complement each other so that the resulting knowledge is of a form that is different than would be possible of any individual engaged in the activity alone. This mutual building of knowledge is often termed *social constructivism*.

Combined with the need to keep up with changes in society, work, technology, and learning theory is the perceived need to keep up with the rest of the world. Despite press that may suggest otherwise, we seem to be teaching mathematics in the United States as well as we always have; we are doing a fine job of teaching our students the skills required for performing as a shopkeeper in the latter half of the 19th and first half of the 20th centuries. However, the rules have changed, and the rest of the world is playing a different game (McKnight et al., 1987). To prepare our students for success in the 21st century, we need to teach them a new game—one that places value on teamwork, problem solving, mathematical reasoning, and forging connections between the world of mathematics and the real world of the student.

The word *standards* was chosen as the appropriate term to describe the goals for reforming school mathematics because it has several distinct, complementary meanings. First, the term *standard* implies that the ideas represented are goals—in other words, the goals outlined in the *Standards* documents represent a standard to which we, as mathematics educators, aspire. The *Standards* represent a vision, an icon that guides our behavior and helps us determine if we are on the right track educationally. Second, the term *standard* implies a level of competence we expect all school mathematics to meet. We have set these standards, and we will accept no materials, teaching methods, or curriculum guidelines that do not meet them. In this sense, the *Standards* are meant as a measure of comparison to determine the worth or value of educational activities. Third, the term *standard* represents a rallying point, a set of ideals that unite us as mathematics educators despite our individual differences in beliefs, knowledge, and teaching styles.

The *Curriculum and Evaluation Standards for School Mathematics*

The *Curriculum and Evaluation Standards for School Mathematics* (NCTM, 1989) outline the content of students' mathematical experiences envisioned to meet the changes discussed in the previous section and describe processes by which these experiences can be assessed. More specifically, the first set of standards in the document, the *Curriculum Standards*, presents four general process goals that are meant to infuse and undergird all aspects of students' mathematical activity. These four process goals are

1. Mathematics as problem solving,

2. Mathematics as communication,

3. Mathematics as reasoning,

4. Mathematical connections.

Notice that these do not represent distinct categories. Instead, any meaningful activity in which a child is engaged should involve a problem situation (one that has no immediate solution process, that requires more than one step to complete, and that extends the child's current thinking); should demand critical reasoning; should make connections to the prior knowledge of the student, to a realistic context, and to other mathematical domains; and should allow for conjecture, debate, and other meaningful discourse among the students.

Taking into account both the general structure of current school districts and the developmental appropriateness of academic tasks, the *Curriculum Standards* outline aspects of mathematical content that should be emphasized in Grades K–4, Grades 5–8, and Grades 9–12. In general, these content standards emphasize number sense over complex paper and pencil computations, the meaning behind arithmetic and algebraic operations over symbolic manipulation, spatial sense and geometric relationships over classification and proof, appropriate estimation and approximation over exact computation, and increased attention to statistical description and analysis over memorizing formulas. The value of the content standards is that they outline a mathematics for *all* students that is significantly deeper, conceptually richer, and more applicable than the mathematics we currently offer that reaches only the elite few.

The second set of standards, the *Evaluation Standards*, describes processes by which teachers and administrators can determine if they are meeting the suggestions made in the *Curriculum Standards*. These evaluation goals describe what assessment should look like in general, what student assessment should entail, and how programs should be evaluated. In general, these standards emphasize that the primary purpose of assessing students is to improve instruction—not to assign grades or check on progress. Assessment should be designed such that it helps teachers understand what students know so the teachers can meet the instructional needs of students better. To do this, teachers must be flexible and design different techniques for finding out what students know. Quizzes and exams are only two narrow definitions of assessment. Observation, interview, and running records such as portfolios are all viable methods for assessing students' knowledge. But the key to implementing the *Evaluation Standards* is determining how each method helps teachers design activities (questions, demonstrations, etc.) that allow students to build their own knowledge in a mathematically powerful manner.

The *Evaluation Standards* suggest that all aspects of mathematical knowledge and its connections should be assessed.

This seems to mean that it is not enough merely to find out what students *can do*, but also how they feel and how they see themselves in relation to mathematics. The ways in which students solve problems are far more important to quality instruction than the answers they produce. Information on students' problem solving, communication, reasoning, connections, and dispositions (substitute the terms *beliefs* and *feelings*), all help us to understand our students' mathematical processes better and help us to teach them better.

Finally, in evaluating a program, it is just as important to understand the nature of the curriculum as it is to critique the methods in which the curriculum is translated by teachers in their classrooms. The nature of the curriculum will put different constraints on what a teacher is able to do with the materials. But also, the ways in which a teacher interprets, modifies, and adapts curricular materials can bolster a weak curriculum or hinder a strong framework. Understanding how the curriculum is perceived by teachers can lead to important changes in both material acquisition and teaching practice.

The *Professional Standards for Teaching Mathematics*

The *Professional Standards for Teaching Mathematics* (NCTM, 1991) describes, with the use of examples and vignettes, how teachers can begin to implement the *Curriculum and Evaluation Standards*. The fundamental assumption behind the creation of the *Teaching Standards* is that teachers are the primary agents for change in schools because they have the most direct access to, and impact on, students. Curriculum, even that emphasized in the *Curriculum and Evaluation Standards*, remains inert without the creative interpretation of a knowledgeable teacher. So the authors of the *Teaching Standards* recognized that the vision of the *Curriculum and Evaluation Standards* would never be realized without a direct example of how such standards impact on teachers' practices and without a method for supporting teachers in their long-term effort of reform.

The shifts in teaching practice advocated by the *Teaching Standards* are major departures from the traditional lecture–drill format so common in American classrooms. The *Teaching Standards* call for classrooms to interact more like communities—that is, an environment in which individuals work together to solve important problems that impact directly and indirectly on their own lives. The strengths of each individual are valued, and the combined input of the class results in a higher level of thought than if each individual worked alone.

As in any community, conflicts will arise as a result of diverse perspectives, opinions, and knowledge. The *Teaching Standards* advocate that logic and mathematical evidence be used to verify mathematical conjectures, rather than relying on the teacher or the textbook as the primary authorities for right answers. This implies that students reason together, solve problems, and invent mathematical models and procedures instead of memorizing preexisting algorithms. Finally, those individuals involved in teaching mathematics should strongly emphasize the connections among mathematical ideas and the connections between mathematics and its applications, so that mathematics becomes viewed as an important way of thinking, as well as an important tool for solving problems.

Six standards for teaching mathematics are outlined in the *Teaching Standards*:

1. Worthwhile mathematical tasks,

2. The teacher's role in discourse,

3. The students' role in discourse,

4. Tools for enhancing discourse,

5. The learning environment,

6. Analysis of teaching and learning.

Like the *Curriculum and Evaluation Standards*, the *Teaching Standards* describe a framework by which teaching mathematics can be assessed and evaluated. Because appropriate assessment should always inform instruction, the *Teach-*

ing Standards also present standards for staff development and support of teachers who are struggling through this significant change.

The *Assessment Standards for School Mathematics*

Although assessment is described briefly and in general terms in the *Curriculum and Evaluation Standards*, and some fine examples of assessment are presented in the vignettes of the *Professional Standards*, one of the most common concerns of teachers struggling with implementing these goals continues to be, "*How* do I assess what these students know? I know *what* I'm supposed to do, but I don't know how to do it." In addition, most commonly used assessment techniques (i.e., standardized tests, state and national assessments, chapter tests, quizzes, and multiple-choice exams) do not provide students with the opportunity to demonstrate their mathematical power, certainly not in the ways envisioned by the *Standards*. It soon became clear to the NCTM that a more detailed complement to the *Evaluation Standards* needed to be produced. A draft of the *Assessment Standards for School Mathematics* (NCTM, 1995) was developed over the summer of 1993 and was distributed to members of the NCTM for review and suggestions. The final draft of the *Assessment Standards* has been published recently and is available to all teachers. All six of the following standards apply to all mathematics assessments, from informal observations and conversations with students to school-wide, district-wide, and state and national assessments and are intended to assist in developing powerful, consistent assessment practices at all levels of information gathering. Assessment should

1. Reflect the mathematics that all students need to know and be able to do,

2. Enhance mathematics learning,

3. Promote equity,

4. Be an open process,

5. Promote valid inferences about mathematics learning,

6. Be a coherent process.

This set of standards is intended to provide a framework for developing new models of assessment that better reflect the vision of the earlier *Standards* documents.

If you have not yet read the *Standards* documents, we suggest that you obtain a copy of each of them for your professional library. The educational and political worlds have embraced the recommendations made in these books, and they truly have become the standard by which curricula, teaching, and programs are evaluated. All teachers of mathematics need to be familiar with the content of these documents and be able to demonstrate how their own practices meet the recommendations. For those of you who are online, you can view a copy of the *Curriculum and Evaluation Standards* from the Eisenhower National Clearinghouse at http://www.enc.org/cd/NCTM/280dtoc1.html. For a hard copy of each of the *Standards*, contact the NCTM:
P.O. Box 25405
Richmond, VA 23286-8161
(703) 620-9840

Statement of Rationale and Goals

Though brief, these descriptions of the major standards for reform should provide a knowledge base for working through the goals of this book. Our rationale for organizing the book around the perspective of a teacher is that we believe that teachers are the primary agents of change in our educational system. Now, with the widespread acceptance of the NCTM standards, teachers have the professional mandate empowering them to take charge and effect change.

Rationale

We had four reasons for writing this book. First, we are providing a medium by which you can develop a philosophical referent for implementing the *Standards* in your own classrooms. Despite the use of illustrative vignettes, examples, and cases, we realize that the reality of your classrooms will be infinitely more complex than we can present. In addition, by their very nature, textual documents are static and cannot fully reflect the dynamic nature of professional change. By providing stimulating examples, thought experiments, and activities, we hope to aid you in modifying and building your own philosophy of teaching mathematics that will serve you when you enter a classroom situation you may never have encountered before.

Second, because philosophy is often so much wind, we will describe practical aspects of implementing reform organized around the three major themes in the *Standards*: curriculum, teaching, and assessment. The key issues related to these themes will be illustrated with examples of how real teachers have dealt with change in their own classrooms. We will illustrate reality here—not perfection. Teachers will fail often, succeed on occasion, and discuss frankly the processes by which they are changing their knowledge, attitudes, and behaviors related to teaching mathematics.

Third, as you become more familiar with the change process, we will embed goal-setting activities in the discussion. These activities are an attempt to stimulate you to examine the avenues available for implementing change and also to recognize the barriers to change you may be asked to overcome.

These sections, called *Self-Directed Questions* or *Activities*, appear at the end of the first four goals and are designed to pull some of these concerns and questions together to assist you in creating a plan of action—to allow you to construct your own solutions to your own problems. They are organized around the primary *Standards* in each of the three documents produced by the NCTM. We will present a vignette or short discussion illustrating a common dilemma in each *Standard*, and then we will ask you a series of questions to help you organize your thoughts.

Do not try to reform all of your practice at once. This often proves to be overwhelming, exhausting, and extraordinarily time consuming. Begin where you feel you can make the most headway in one or two areas. Once you begin the process of change in these areas, more questions will arise that will lead you to begin change in the other areas. In other words, as you work through these questions, do not go through each of the recommendations in the *Standards* in order, but start where you feel you need the most help and create a workable set of goals for beginning the change process.

The optimal situation for creating professional goals is one in which groups of teachers plan together. Thus, these activities are designed to be interactive, involving pairs or groups of teachers working toward the same end. Get a partner—another teacher in your grade level or at your school will have many of the same needs as you have and can serve as a primary source of support as you both struggle. Work through the questions together, clarifying ideas, setting mutual goals, and searching for a common understanding of the ways in which you can reform mathematics instruction as a team. We have attempted to eliminate unnecessary jargon from our discussion. However, if you find a technical term with which you are unfamiliar, or if you want to know how we conceive of a particular term, we have provided a glossary for your reference.

Fourth, because no change can be attempted without resistance (both within ourselves and from others), in the final goal, we will discuss support mechanisms that will foster realistic development toward meeting your professional goals.

A short discussion of technological issues appears in goal 5, but we do not deal extensively with this topic. The radical differences in the level of technology currently available at different schools makes this discussion too long for the format of this book. However, we do feel that technologies are vital to teach mathematics effectively in the modern classroom. For a theoretical discussion of technology as a tool for thinking, see Pea (1987). For a practical discussion of technology in the classroom, refer to Jensen and Williams (1993).

As you work through the book—looking for hints to implement the standards that have been set for mathematics teaching, learning, and assessment—you probably will be somewhat disappointed that no hard-and-fast rules exist for creating the kind of learning atmosphere envisioned by the NCTM. What currently exist are suggestions, rules of thumb, and research-based methods that have worked in other classrooms. Our intent is not to provide methods for you, but to raise questions and concerns to get you thinking about how you might begin to reform your own teaching.

Goals

The goals of this book are to

1. Help you build a new understanding of the nature of mathematics and mathematics teaching and learning,

2. Help you build and choose mathematically powerful curricula to promote students' problem solving,

3. Help you build new teaching strategies to facilitate students' mathematical problem solving,

4. Help you develop a balanced set of assessment strategies to uncover students' mathematical understanding,

5. Provide you with a list of practical resources to help you begin and sustain your change process.

goal one

Building a New Understanding of the Nature of Mathematics

To set the stage for understanding reform from a teacher's perspective—for developing and improving our ways of teaching mathematics—we need to revisit or refresh our memories with the methods by which we were taught math. In a traditional mathematics class, the teacher spends the first few minutes of the class with some type of review, possibly in the form of correcting the previous night's homework. In the second part of the lesson, the teacher introduces a new concept or procedure—primarily with the teacher as a lecturer and the students in the listening role, perhaps taking notes. This is followed by a few practice problems, and then in the final 15 or 20 minutes, the members of the class continue practicing on their own a set of problems from the textbook (either the even or the odd numbered problems, whichever does not have the answers given in the back of the book) or a work sheet.

As you may recall, most of the math we learned as students (in terms of the amount of time allocated) was centered around computation. Students who had the tool of memorization down pretty well were probably successful at math. But even the successful students were often thrown by story problems. Those agonizing story problems caused many a headache. They did not fit the typical book problem and they occurred only once in a while at the end of a lesson. Some of these problems became so frightening they were given special names:

> The Train Problem: If a train left Philadelphia at 10:00 going 57 miles per hour, and another train left Washington, D.C., at the same time, where would the two meet and at what time?
>
> The Bathtub Problem: A bathtub can be filled by its faucet in 5 minutes. The drain can empty the bathtub in 7.5 minutes. If the faucet is left running, and the drain is left open, how long will it take to fill the tub?

To make matters worse, the evidence accepted for success on math problems was having the exact answer that was in the teacher's book. Unless students enjoyed fooling with numbers, the structures of thought, procedure, and verification were so set that the students became bored, and they soon disengaged.

From the perspective of teachers, the pace of instruction was started at chapter 1 in the textbook, "Whole numbers and place-value," and the goal was to see which teacher could get farthest in the text by the end of the last grading period. Quite often, those "fun" chapters at the end of the book were never gotten to because we often got bogged down in decimal operations or fractions. Geometry, integers, volume, statistics, probability . . . these chapters were either not touched or we were allowed just a brief encounter.

Often in our teaching we have sensed the struggle of our students to understand. The authors of this book are no exception. We have asked ourselves often, "How can we engage students in their learning?" Many times it seemed as if we were dragging kids through the chapters, and when questions like, "When are we ever going to need this?" "When will we ever use this?" came up, we often felt a little uneasy, because often our answers really did not satisfy either our students or ourselves. Many of us realized that something needed to be changed, but how?

A VIGNETTE

The vignette that follows illustrates some of the changes that are expected to take place in the mathematics classroom as a result of implementing the NCTM *Standards*. Imagine yourself as an anthropologist studying the beliefs about learning evidenced by a certain culture. You have entered into a 6th-grade classroom and taken the role of a student to discover what learning mathematics means to this culture. You are sitting in a group of 4 students, and you notice that there are 9 groups in the class, making a total of 36 students. The teacher begins the lesson by wheeling in a television monitor and VCR. She instructs the class to view a film clip from the movie *Rosencrantz and Guildenstern Are Dead.* In the clip, Rosencrantz and Guildenstern, two travelers, find a gold coin. Guildenstern flips the coin and announces, "Heads!" Not content with one trial, he flips the coin again, "Heads!" Rosencrantz watches him contemplatively as he continues to flip the coin, "Heads . . . heads . . . heads!" Exasperated, Rosencrantz seizes the coin and examines both sides. It turns out to be a fair coin. They continue their travels with Guildenstern flipping a total of 78 heads in a row. At last Rosencrantz takes a long look at the coin and announces, "I feel a change about to happen," and flips the coin: heads. "Well," he muses, "it was an even chance."

The teacher turns off the VCR and asks the class, "I noticed that some of you were giggling at certain points in the clip. I want you all to get into your groups and answer this question, 'What makes the clip funny?'"

The question takes both you and the other members of your group off guard. What? You have never thought that humor could be a mathematical question. Marly, sitting across from you, ventures forth a first idea for discussion, "Well, I was thinking, 'This is impossible. You would never get 78 heads in a row.' "

The teacher, overhearing, runs up and asks, "How do you know? Is it really impossible?"

"Well, it may be *possible*, but I wouldn't bet on it," Jaime states confidently.

"Why?"

"It's like, 50–50. You should get as many heads as tails."

"How do you know that? Would you be surprised if I flipped the coin twice and came up with heads both times?"

"No," says Carole. "But I *would* be surprised if you got 78 heads in a row like in the movie. It should *average out* to be 50–50." (A chorus of "yeah" from the other members of the group.)

"Okay. Come up with a method for testing that assertion. When I come back you can show me if your idea is correct or not."

The teacher then leaves and listens to other groups of students struggling with the problem. All are asked to come up with a similar method. A short whole-class discussion allows for some tailoring of methods so that results can be combined. You work with your group, flipping coins and recording the proportion of heads and tails. You never actually get 50% heads, but you begin to notice that as the number of flips gets larger, the proportions slowly get closer to 50–50.

The next day, in a whole-class discussion, each group combines its results, and the frequencies are graphed on the board. Carole presents her hypothesis and states, "See, the average number of heads should be right there in the middle of the graph. That's about half of the total number of flips, so the number of heads and tails is about 50–50."

"Can we check Carole's reasoning another way?" asks the teacher.

"Well, I used my calculator and found the average by adding up all the groups' data and dividing by the number of groups," says Michael. "I got an average of 49.2 heads out of 100 tosses per group. That's pretty close to 50%."

"Good. You can all see these are two ways to think about averages. The center of a distribution like we see on this graph and the arithmetic way—adding up the scores and dividing by the number of scores."

SOCIAL CONSTRUCTIVISM: KNOWING AND DOING

The nature of learning mathematics in the previous vignette is assumed to be quite different than the ways in which you probably learned mathematics. Earlier models of human learning likened the student to a large sponge, absorbing the content told by the teacher or illustrated in the textbook. Like a sponge, information could be squeezed from students from time to time to determine if they had absorbed enough. These periodic squeezings are still called exams, tests, or quizzes (NCTM, 1970).

Later, findings from educational psychology suggested that knowledge is built up bit by bit, from little pieces into a larger hierarchy of knowledge (Ausubel, 1977). To teach effectively, teachers were expected to determine the correct sequence of tasks (the little bits), and order them in such a way that they became a coherent set of skills mastered. Although intuitive on the surface, this approach has proven to be problematic. Often the little bits became so isolated from the larger content structure and from the world of the child that students had extreme difficulty seeing the connections in the ways their teachers had in mind.

Current research suggests that students do not absorb knowledge passively, nor do they accrue knowledge bit by bit. Rather, students *actively build* their own knowledge, searching for new information, determining how it fits their understanding and how it does not, and making appropriate modifications to their thinking to accommodate the new information (see Romberg & Carpenter, 1986). This new understanding of learning is termed *constructivism* (Steffe, 1988; von Glasersfeld, 1985) In this philosophy, the learning process is likened to a construction site, where information serves as the building material, content functions as the blueprint, and the student is both the designer and builder. This basic philosophical orientation is fundamental to understanding the *Standards*.

Mathematical thinking is fostered through social interaction. Argument, debate, and refutation allow students to see diverse ways of solving mathematical problems and lead them to an autonomy and sense of self-efficacy with respect to mathematics. All knowledge exists as *interpretations* of information existing in the world of the child (see Wertch, 1985, for a detailed explication of interpersonal interpretations of information). Because human beings in a given culture have similar experiences, their interpretations tend to be compatible. However, no two persons have exactly the same background and experience, so their personal interpretations (or *knowledge*, if you will) of any situation will differ slightly. One only has to give a cursory glance at our political system to see vastly different understandings developed from the same information source. The same holds true for mathematical information. In any mathematics classroom of 30 children, there will be 31 different interpretations of the concepts and procedures being taught (the teacher also interprets mathematical information in a unique way). It is often the discussion arising from differing views on mathematics that leads children to higher levels of understanding.

Looking back again to the structure of the math classes we probably took as students, how many of us remember working cooperatively with other students? Chances are, we cannot remember any, or even if we can remember working with others, it was probably in isolated instances. If we look at the workplace, how many jobs require working in isolation? In reality, most positions require employees working together, interacting, tackling a big job with a joint effort. When individuals do work alone, we possibly work on a small piece of a really big problem and then regroup with other members of our team to finalize the task. Problem solving, communication, reasoning, and connecting ideas are commonplace in the working world. The mathematics we engage students in must reflect these workplace values.

But argument, debate, and the construction of knowledge are not fostered without some purpose. Routine tasks stifle thought and lead to complacency because they do not serve children's natural drive to make sense of their world. The *Curriculum Standards* make the following assertion:

> Three features of mathematics are imbedded in the Standards. First, "knowing" mathematics is "doing" mathematics. A person gathers, discovers, or creates knowledge in the course of some activity having a purpose. This active process is different from mastering concepts and procedures. We do not assert that informational knowledge has no value, only that *its value lies in the extent to which it is useful in the course of some purposeful activity*. (NCTM, 1989, p. 7, italics ours)

goal two

Building and Choosing Curricula

A primary focus of the call for reform is to change the nature of the tasks we involve our students in. As a result, the large burden of developing new lessons, and selecting appropriate materials is left up to teachers. This goal helps you develop your curriculum to be more attuned to how children think mathematically, resulting in a more coherent and useful structure to their knowledge.

A *TRADITIONAL* APPROACH TO VOLUME

We have all experienced *the* formula for volume:

$v = l \times w \times h$

where *l*, *w*, and *h* are symbols standing for the length, width, and height of a rectangular prism. Usually, before we begin work on volume, we study area (length x width). Our students generally understand how to compute area using this formula, and volume is a natural extension by adding the new dimension of height. Typical textbook problems that appear following work on area look like this:

The dimensions of a rectangular solid are given as:
  Length: 4
  Width: 5
  Height: 6
Compute the area of the solid.

The intention of traditional curricula is that after practice with these types of problems, our students will not only know how to do the problems and come up with the correct answer, but that they will also understand the concept of volume. Unfortunately, being able to compute the volume of a box with given dimensions does not lend much insight into techniques for determining volume of nonrectangular solids (spheres, cylinders, irregular figures). Moreover, because students' knowledge is so situated in this specific situation (e.g., Brown, Collins, & Duguid, 1989), it is unlikely that the general notion of volume as "filling space" will be developed by the students without considerable experience of a different sort.

A *CONTEXT* APPROACH TO VOLUME

From the journal of the second author:

> A few years ago during the spring, I could tell the students were getting bored with the same old thing and needed a little renewal. I had had some training in an

innovative curriculum project but was far from seeing the total picture. I had to do something to spark interest in my students. As they were staring out across their desks, I looked at the textbook pages on volume and struggled for enlightenment.

"How many dimes will fit in your desk?"

At my school, we have the type of desks that hold books in a compartment. Most desks today are not like that, but you could adjust this problem by using boxes (all different sizes would be preferable). I had my students in flexible groups of about 2 to 4 students. This spur-of-the-moment question developed into a whole scope of mathematical concepts during my students' discussions: measurement and converting measurement units, estimation, money and money units, and calculation. I probed students with *how* and *why* questions to get at their reasoning strategies. As students searched for more efficient ways of finding the answer than physically filling their desks with dimes, they developed various versions of the standard formula for rectangular solids: $v = l \times w \times h$.

One point that must be made is when we use this type of question, the main concept (in this case volume) is not isolated and is not the total focus of the lesson. Rather, it is mixed in and connected with many other concepts. What began as a one class-period question has now developed into a three-day project in which students compare their class results with that of other classes, finding averages for all five class periods, determining the amount of money that could be stored in all 32 desks in my room, and graphing the results on the computer. A good question can go a long way!

THE NATURE OF MATHEMATICAL ACTIVITY

One of the biggest hurdles in changing our methods of teaching mathematics is creating appropriate mathematical experiences for our students. The current curricula available from textbook publishers is changing, but they have not yet reached a stage where they could be considered to *embody* the philosophical intent of the NCTM *Standards*. Replacement units that do reflect the intent of the *Standards* help, but as of yet, they do not form a cohesive set of experiences around which we can organize an entire year of instruction. At present, to make our teaching better reflect the vision of the *Standards*, we must be the ones to gather, develop, modify, and organize curricular materials into a set of experiences around which our students can construct mathematical meanings more easily. If this is the case, what should we look for? What should we do?

As a general rule of thumb, mathematical curricula should be *active*—that is, it should require the mental (and often times physical) involvement of the students. This heuristic has been useful for many teachers. Search for activities that engage students' intellect, get them discussing difficult ideas and developing models and other representations to communicate their thinking. However, being active is only one facet of good mathematical activity.

A second feature of mathematical curricula that is often overlooked in the search for fun activities is that mathematical tasks should lead somewhere. Children should not be expected to create or invent the entire history of mathematics on their own regardless of all the rich experiences we can dream up. Rather, they should be guided through this process by involving them in cleverly designed activities that tie together in important conceptual ways. For example, in the activity described previously, the students developed a beginning concept of area and volume. However, the formulas they created are appropriate only for rectangular solids. What about spherical shapes? What about really odd containers like ponds? Continuation of the original problem to more complex problem situations allows students to expand their understanding of volume and connect it with other geometric concepts dealing with shape.

PRINCIPLES FOR CHOOSING AND DESIGNING CURRICULA

The following principles for choosing and designing curricula are a compilation of suggestions and research findings from a number of sources. Romberg and Tufte (1987) deserve the most credit for providing the general structure and the metaphor of curricula as story. Mandler's (1984) and Schank's (1990) discussions of how humans process information in story form also contributed mightily to the metaphor of mathematical activity as a story with themes, characters, and conflicts.

Principle 1: Mathematical Curricula Should Be Realistic

Learning mathematics involves the *construction* of knowledge into a *structured* entity. New concepts are fit into the child's existing knowledge structure or trigger the modification of their knowledge to accommodate the new information (Treffers, 1991). For the purposes of instruction, this means that the concepts we want students to develop must be identified and a curriculum built around these concepts to emphasize their relations to other mathematical and nonmathematical concepts. These domains (counting and numeration, measurement, geometry, etc.) should not be considered independent of each other. Rather, their unique contributions to the field of mathematics should be emphasized in relation to what they add to the students' understanding of their world.

Each curricular unit (about 2 to 4 weeks of work) should introduce or reintroduce these mathematical concepts to the students, and there should be a new problem situation to be resolved that involves some kind of conflict, suspense, crisis, and resolution similar to the theme of a story. Students should be expected to construct meanings, interrelate concepts and skills, and use those meanings in a variety of problem situations. The activities you design may involve a story about the world around the student, scientific discoveries, historical events, or you may decide to discuss mathematical *characters* themselves.

In general, curricula should allow students to represent quantitative and spatial relationships in a broad range of situations, communicate those relationships using the language of mathematics, make appropriate use of technology to carry out numerical calculations, make predictions about relationships, and interpret the results of the investigation.

Principle 2: Mathematical Curricula Should Provide Access at a Number of Cognitive Levels

Because thinking about a mathematical phenomenon can occur on a number of levels of abstraction (by which we mean the shift from dependence on physical models or concrete experiences to more formal systems of thought), curricula should be structured to provide each student access at a number of given levels. These multiple access routes can be made possible by using mathematical models that embody the underlying mathematical concept being developed. For example, Streefland (1986) described the use of fair sharing as a model for teaching fractions. Students begin with real objects, cutting them and sharing them equally among themselves. As the students come to understand the part–whole relationship of fractions, they can progress to the use of graphs, common fractions, or other more abstract models that illustrate the concept of ratio and equivalence of fractions (see Middleton & van den Heuvel-Panhuizen, 1995). Students with only informal knowledge of fractions may start with the fair sharing model. Students who understand partitioning and part–whole relationships prior to the lesson may start with the other models. These different access points should not only occur in the building up of a concept over a series of units. Very often it is possible to embed them within a single chapter or even within a single exercise. In this way, students beginning to understand fractions can use their models to communicate with other students who may be using more sophisticated models.

For each new activity students should be able to make

a new choice about at which level to start the climb, according to their self-perceptions of ability. This provides avenues for success for all students and allows students to exercise a certain degree of free choice, both essential components for developing interest and achievement motivation for the activity (Middleton, 1995; Middleton, Littlefield, & Lehrer, 1992; see also McCombs & Pope, 1994).

Ideas are best introduced when students see a need or a reason for their use. To be effective, events in the unit must be *causally* and *hierarchically* linked. That is, as new problem situations arise, the information required to solve the problem should stem from the natural world of the students so that they can attach the mathematics to their existing knowledge structures, and the problem should extend this knowledge in such a way that the mathematics becomes another way of conceptualizing the real problem—in other words, the *theme* of the lesson.

Students construct their mathematical knowledge through interaction with activities that are (un)meaningful. Through the use of rich contexts, full of mathematical relationships, students see patterns in the realistic situation that require numerical or spatial processing to understand. In all cases, students will fit the new experiences with their existing cognitive structure, modifying the structure as needed. The trick is designing a set of experiences that will be both useful practically and coherent mathematically.

In designing curricula, the base level for any activity should require either little prior knowledge or knowledge that is easily retrievable to the student to ensure that all students in the class will be able to approach the problem at hand. As students progress in their knowledge and discrepant information, a greater variety of problem situations or more difficult applications may be introduced to convince the student of the generality of the solution.

Principle 3: Mathematical Content Should Reflect the Prior Knowledge of the Student

All curricular sequences need to be adapted and modified in light of the knowledge the students bring to the unit and the context in which instruction takes place. This is one of the fundamental ways in which we can begin to implement the *Standards*. We teachers know our students. We know what they like and what they dislike. We have a good idea what their prior experiences have been, and we can make informed judgments about how to build on their prior knowledge. This can be a tricky balancing act. As in a mystery novel, enough information must be made available for the students to be able to come up with a method for solving the problems, but enough information must be withheld so that the students' solution strategies are meaningful extensions of what they already know. In a mystery, the detective must make certain assumptions regarding the motives, opportunity, and character of each of the prospective suspects. In a mathematics unit, students also need to make certain assumptions about the nature of the problem, the efficiency of different strategies, the pattern of data, and piece this information together to come up with a reasonable facsimile of the problem and possible avenues to resolve it. This type of problem solving requires that students have an opportunity at key points in the curriculum unit to reflect on what they have learned and to project what might lie ahead (Treffers, 1991).

It is often the restrictions in data and methods that define the boundaries of a mathematical concept. In other words, the students need to develop, from their prior knowledge and experience, a theory about the nature of the mathematical entity they are studying, and they must be given ways in which they can test the adequacy of their theory so that the structure of the mathematics becomes more apparent. If we look on mathematics as a part of an integrated educational system, these restrictions and boundaries are of fundamental importance.

Principle 4: Mathematical Curricula Should be Interactive

As indicated previously, we are concerned not only that students be able to make sense of the presented mathematical material, but we are also concerned that they be able to create their own methods and operations for the purpose of making sense of their world. For this reason, we suggest creating many instances in the lesson where the answers to questions are not obvious, where they can be approached from a wide variety of viewpoints and strategies, or where they require the integration of multiple mathematical and cognitive representations. At certain key junctures in a curriculum unit, there must be the possibility for the student to explore "tangents" that they find interesting, important, and ultimately meaningful in a personal sense.

Moreover, as we discussed earlier, the use of mathematics in the real world is not usually performed by the individual isolated from others. It is something that involves a group of individuals working toward some common goal (remember *social constructivism*?). The unique perspective and expertise of each person in a working group adds to the total knowledge and creativity of the group, increasing the power of the mathematics being examined and the ability of the student to communicate findings to a wider audience. Thus, the opportunity for students to work together on larger projects that require discussion, argument, and compromise is crucial in their preparation for the world. Examine the curricula you are using currently. Is communication valued and built into the types of problems students are expected to solve, or is it placed in *Enrichment Exercises* or *Extensions*?

In addition to mathematical content, the units also must stress flexibility and adaptation. The needs of business, industry, and science dictate that individuals be able to adapt to a work environment that changes as new technology develops, as the needs of the consumer shift, and as global competition grows more intense.

THE ROLE OF MODELS IN TEACHING MATHEMATICAL CONCEPTS

At this time, we would like to pay closer attention to the use of models to teach mathematical concepts. Models are ways of representing mathematical ideas that make the concepts more apparent to the students. Manipulatives, graphs, tables, analogies and metaphors, and stories all can be used as models for illustrating important facets of mathematical topics. But models can play a much deeper role in instruction than serving as mere illustrations. Models can serve to embed the mathematical concepts being learned in the reality (both mental and physical) of the student. Developmental psychologists believe that children begin learning in a fundamentally *concrete* way (Siegler, 1986). That is, they learn by doing. They take things apart and attempt to put them back together again to see how they work. They arrange things that are similar into groups and then rearrange them based on other attributes. For children's thinking to develop appropriately, it is crucial that they be able to do these things. We learn counting by counting *things*. Without *things* to count, the meanings and relationships among numbers cannot be internalized. It is only after a large number of experiences of counting objects that understanding of numbers begins to take a more abstract (i.e., internalized, generalized) form.

This modeling is especially crucial for the mathematical development of young children. Carpenter, Ansell, Franke, Fennema, and Weisbeck (1993), for example, exposed 70 kindergarten children to nine word problems, ranging from simple subtraction to nonroutine division. They found half of the children were able to use a valid strategy for solving *all* of the problems. About two thirds correctly solved seven of the nine problems, and nearly 90% used a valid strategy for solving the basic subtraction and multiplication problems. In addition, the ways in which the children solved the problems were interesting. All of the children directly modeled the actions implied in each problem.

In Table 1, the 11 different problems that can be created by combining the *actions* implied by the story context and the unknown quantity (start, change, result) for simple addition and subtraction are presented. In problem 1, for instance, the typical solution strategy of a kindergartner would be to first count up five objects (unifix cubes, counting bears, marbles) and put them in a pile. Next, she would count up a second set of eight counters and put them in a separate pile. She would then *join*

Table 1 Addition and Subtraction Problem Types

Problem Types	Problem Wording
1. Join (result unknown)	Jamie had 5 marbles. Mario gave her 8 more marbles. How many marbles does Jamie have altogether?
2. Join (change unknown)	Mamphono has 6 pencils. How many more pencils does she need to have 10 pencils altogether?
3. Join (start unknown)	Lee had some candles. Mary gave him 7 more candles. Now he has 12 candles. How many candles did Lee have to start with?
4. Separate (result unknown)	Tracy had 14 donuts. She gave 8 donuts to Julia. How many donuts does she have left?
5. Separate (change unknown)	Nick had 13 cans of creamed spinach. He gave some to Juan. Now he has 5 cans left. How many cans of creamed spinach did Nick give to Juan?
6. Separate (start unknown)	Beth had some cherries. She gave 9 cherries to Stephanie. Now she has 3 cherries left. How many cherries did Beth have to start with?
7. Part–part–whole (whole unknown)	Consuelo has 7 red marbles and 4 blue marbles. How many marbles does she have?
8. Part–part–whole (part unknown)	Herb has 9 agave plants. Five plants are small and the rest are large. How many large agave plants does Herb have?
9. Compare (difference unknown)	Dave has 9 water balloons. Margaret has 5 water balloons. How many more water balloons does Dave have than Margaret?
10. Compare (compare quantity unknown)	Karen has 9 pieces of candy. Jon has 5 more than Karen. How many pieces of candy does Jon have?
11. Compare (referent unknown)	Alfinio has 9 strawberries. He has 5 more strawberries than Mary. How many strawberries does Mary have?

Note. Adapted from Fennema & Carpenter (1994).

the two piles and count all of the objects to obtain the answer 13 marbles. Notice that the actions of the child parallel the actions implied in the written problem. At the earliest stage of arithmetic understanding, children attend to the literal interpretations of word problems and attempt to model the scenarios explicitly. The problems are not seen as *addition* or *subtraction* problems but merely as interesting little mysteries that can be solved using intelligent strategies. Often children will solve a *subtraction problem* (e.g., problem 2) using a strategy that we, as adults, would think more typical of an addition problem. How do you think a typical kindergartner would go about solving problem number 2?

The research literature on young children's solutions of such problems indicate that students begin thinking about these problems by *directly modeling* the situation with counters, tallies, or other ways of keeping track of the numbers in the problem. The most typical strategy would be to count out six objects, and then count out "7, 8, 9, 10," keeping track of the second set by separating it from the first. The second set then would be counted to obtain the answer. The same general principles can be applied to young children's beginning understanding of multiplication and division problems.

These findings are remarkable because they suggest that children are far better at arithmetic far younger than textbook scope and sequence would have us to believe. The majority of children as young as kindergarten can make sense of, understand, and solve successfully, addition, subtraction, multiplication, and division problems *given an appropriate set of mathematical experiences*. Please note that the research also indicates that children can make sense of these problems only if they are put in context. That is, a child who successfully can solve the problem: "Robin has 3 packages of gum. There are 6 pieces of gum in each package. How many pieces of gum does Robin have altogether?" by placing 3 sets of 6 counters out on the desk and counting all 18 objects, will most likely *not* be able to solve the problem "$6 \times 8 = ?$" without a lot of rich experiences making sense out of grouping and repeated addition situations (Carpenter et al., 1993, p. 434).

This leads us to a fundamental postulate of implementing the *Standards*.

> Considerable rich experience modeling mathematical situations must precede and accompany more abstract symbolic manipulation.

In upper elementary and middle grades, students' initial reasoning about fractions often stems from the fair-sharing situations that we described earlier. No one wants to be stuck with the little half of the cookie, so children devise ingenious ways of determining when the portions are the same. Often, you will find children placing one half of a cookie on top of the other half to see if there is any overlap. As a natural part of students' reasoning about proportion, fair-sharing activities serve as a model for basic understanding of fractions. Two cookies shared among five children necessitate dividing each cookie into a number of pieces divisible by 5. Each child would then receive two of the small pieces. In this example, the *model* of the fair-sharing situations embodies the mathematics we wish to teach. Through involvement in these activities, children necessarily engage in proportional reasoning. *Fractions* as numeric symbols with a numerator and denominator then become mere notations that make it easier to communicate the underlying ideas. As students get older, the real cookies can be phased out for pictures of cookies, fraction bars, or other more abstract proportional models. In turn, the fraction notation can be extended and connected to percentage, decimal, and ratio notations (Middleton & van den Heuvel-Panhuizen, 1995).

Post, Cramer, Behr, Lesh, and Harel (1993) maintained that it is the translation between modes of representing rational numbers (i.e., making connections between a picture, a symbol, and a manipulative activity representing 1/2) that makes ideas meaningful for children. For example, take a look at the following problem:

> Susan is the president of Sue and Joe's Cookie Emporium, a two-person company that sells cookies to local supermarkets.

Joseph is the vice president. As president, Susan gets $2 for every $1 Joseph makes. All profit made by the company goes toward salaries.

1. If Susan makes $30,000 in salary this year, how much does Joseph make?

2. What is the total amount of salary paid out to the employees?

3. What percentage of the total salary do Susan and Joseph receive?

4. Imagine Sue and Joe's Cookie Emporium becomes very successful. Susan and Joseph want to be able to figure out their salaries for any amount of profit. Describe a way they can do this.

The first question can be solved in a number of ways. Using manipulatives, students may use 30 counters to represent Susan's $30,000. Then they can pull out sets of two counters and count the number of sets. This will result in 15 sets of two counters, which represents Joe's salary of $15,000. Or a student may choose to create a bar representation (see Figure 1).

An older student might be able to calculate all of the first three solutions directly. But when extended to the fourth part of the problem, the general solution might be more tricky. Here, a line graph might be a useful model (see Figure 2).

These older students apply their understanding of patterns, fractions, percentages, and ratio together to solve the problem. In addition, they are beginning to put together algebraic notions (the idea of function and linear graphs) to obtain the general solution to the problem. Notice that the underlying ratio of 1 to 2 for Joseph's to Susan's salary becomes broadened as students explore the part–part *ratio* of 1:2, the part–whole *fraction* of 1/2, and the *percentage* of 50% simultaneously using the different models, and that this ratio is the slope of the linear function $f(x) = 2x$.

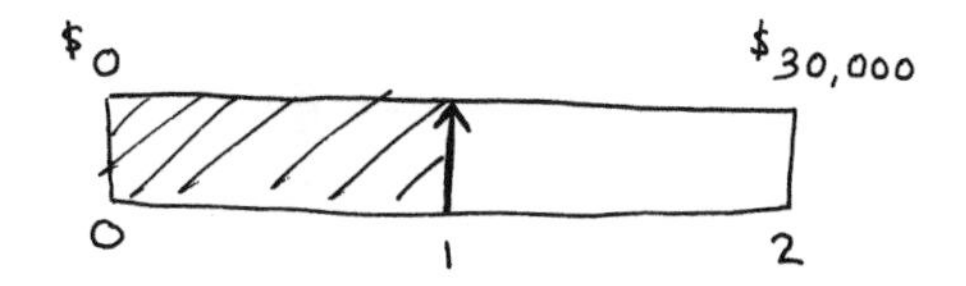

Well, from the bar, 1 is half of 2, so Joseph's Salary is ½ of Susan's. So I put the $30,000 in the calculator and came up with $15,000 for Joseph's salary.

Since Joseph makes $15,000 and Susan makes $30,000 they both together make $45,000.

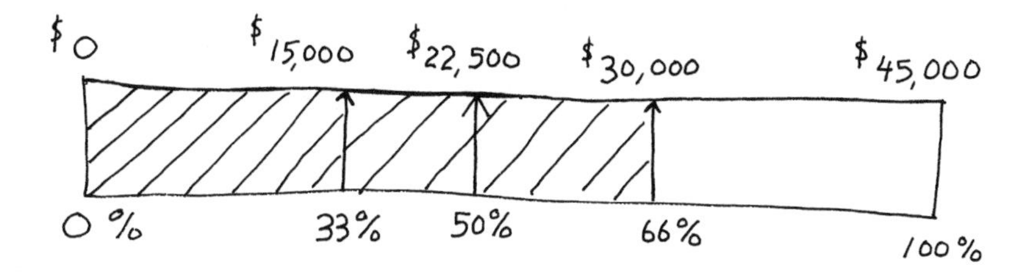

Since Joe only makes ½ of what Sue makes, his salary has to be less than 50% of their total which is ½. So, I drew another bar, and saw that 15,000 was about ⅓ of $45,000, so his percent has to be ⅓ of 100% and that's 33%. Since Sue makes 2 times as much as Joe, she has to make 2 times 33% which is 66% and that is just about $30,000.

For the last question, Sue and Joe could just take any amount of profit, and divide it into 3rds to find Joe's salary, then double that to get Sue's.

Figure 1 *Students' Solution for Problems 1 Through 4 Using a Bar Representation*

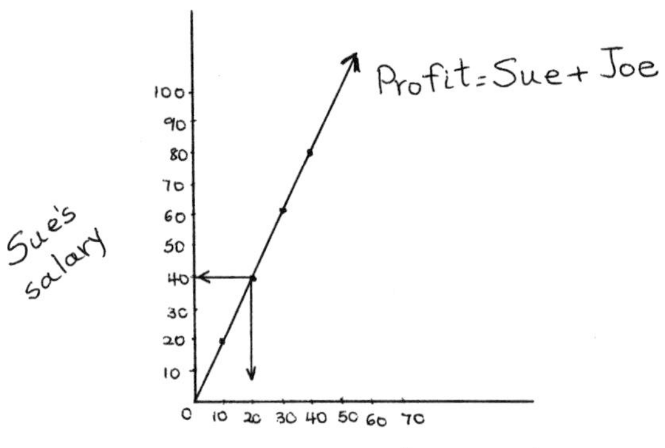

Profit	Sue's salary	Joe's salary
0	0	0
30	20	10
60	40	20
90	60	30
120	80	40

From our graph and table you can see that for every 30,000 dollars they make, Joe gets 10,000 dollars, so if they made 60,000 dollars profit, Joe would get $20,000. If you go over to Sue's axis, you can see that for $60,000 profit, she would get $40,000. You can check it because $20,000 + $40,000 = $60,000. So, for any amount, you could just read their salaries off the graph or figure it out.

Figure 2 *Students' Solution for Problems 1 Through 4 Using a Line Graph Representation*

This translation between table, graph, and formulas (in this case, the formulas are implicit), draws the students' attention to the important similarities between the meanings of different representations of rational numbers.

This leads us to a second fundamental postulate:

> Any single model or representation of a concept is insufficient for developing mathematical power. Whenever possible, encourage the use of multiple models.

Even if students understand a concept and can solve problems in a single powerful way, communicating what they know can be difficult without resorting to multiple models. The old adage, "A picture is worth a thousand words," should be taken seriously in the modern mathematics classroom. Students should be encouraged to communicate their thinking using a variety of appropriate representations including graphs, tables, formulas, verbal rules, and drawings. Other models such as flowcharts, geometric solids, metaphors, and analogies should become commonplace in the mathematics classroom. Please note that including these models does not detract from the focus of the mathematics; it merely highlights and deepens the meanings underlying the desired mathematical concepts.

The *Standards* place a heavy emphasis on drawing connections among mathematical ideas and between mathematical ideas and experiences in the real world. In our experience with preservice teachers, we have found that beginning teachers tend to focus on the latter arena and tend to overlook the former when they evaluate curricula and develop instructional activities. This may be a result of their excitement in finally seeing where the mathematical concepts they have struggled with actually have meaning, but we think it goes deeper than that. It is much easier to involve students in an interesting activity in which they are required to apply mathematical ideas—arithmetic, formulas, and graphing in a statistical investigation, for example—than it is to develop a series of mathematical experiences that are mutually informative and that lead to higher level understanding of the mathematical concepts.

There are distinct practical advantages to examining our curricular frameworks in terms of the intramathematical connections they foster. First, if the recommendations in the *Standards* were implemented as separate and distinct concepts, the time required to cover the material would be staggering. Take the grades 5 through 8 curriculum *Standards*, for example. Under reasoning, the *Standards* lists four areas to receive increased attention: (a) reasoning in spatial contexts, (b) reasoning with proportions, (c) reasoning from graphs, and (d) reasoning inductively and deductively. Receiving decreased attention is only "relying on outside authority" (teacher or an answer key). If these areas of reasoning were to receive equal emphases in instruction—individually and not connectedly—considerable time would be wasted going over remarkably similar material with no emphasis devoted to the ways in which graphs, geometry, and proportional contexts intertwine in real problem solving.

Second, we are concerned not only with teaching mathematical concepts to our students but with developing students' *understanding* and ability to *apply* these concepts. Research in educational psychology tells us that the understanding a person has in a given area has a lot to do with the number of connections that exist between concepts in a person's memory (Anderson, 1990). In fact, some researchers go so far as to say that understanding *is* the pattern of connections a person makes between concepts (McClelland & Rumelhart, 1986). The more connections, the more ways a person can think about a given problem and the more routes the individual will have to develop a solution strategy. Thus, to develop real understanding of mathematical ideas, concepts must be connected to each other. The only way to do this is to develop experiences in which students must bring different mathematical knowledge to bear on a common problem. For example, taken together, geometric concepts of similarity and right-triangle trigonometry can be seen as instances of proportional reasoning. Graphing the ratios involved in such problems provides a visual tool for reasoning about the interrelationships between angles and sides in geometric figures. All of these areas can be reasoned together, providing a more powerful instructional medium for exploration, a more connected knowledge structure (and thus more useful) for students, and because the teaching of these concepts occurs simultaneously, it provides more effective time management for teachers.

Gravemeijer (1994) described five types of activity that fit a developmental model of human mathematical thought (see also Freudenthal, 1983; Van Hiele, 1985). These activity types begin at the exploration stage of knowledge development (this can be interpreted as *concrete*, in the Piagetian sense), and lead through successive levels of abstraction and interpretation to a high-level system of connected content knowledge (this can be interpreted as *formal* in the Piagetian sense, although it does not necessarily translate expressly to the concept of *formal operations*). As such, it makes a fine model for thinking about activity situated in a coherent instructional sequence. These five types of activity are

1. *Phenominological exploration:* Students' activities in a given domain should begin with rich experiences that beg to be organized, and instruction should focus on helping the students organize these experiences in meaningful ways.

2. *Bridging through the use of models:* Models, as described previously, rather than being *given* to the students, should "*arise* from problem-solving activities and subsequently can help to bridge the gap between the intuitive level and the level of subject-matter systematics" (Gravemeijer, 1994, p. 451, italics ours). In other words, models arise from phenominological (or level 1) experience, and help students move to higher levels.

3. *Student contribution:* Students' constructions and productions (their questions for inquiry) orient the teacher to their learning and suggest common ways of interpreting models and experiences.

4. *Interactivity:* Discourse, negotiation, cooperation, and analysis should be used as ways of moving students' informal concrete understandings to more formal, systematic understandings.

5. *Intertwining:* Learning strands should be interconnected so that they cannot be dealt with as separate entities. In this way, students' knowledge becomes interconnected. Mathematical applications are prime ways to intertwine content strands.

In choosing curricula and designing learning experiences, be sure that the aspects discussed in this goal are *built in* appropriately. If these levels of activity are not included in the design of curriculum materials, it is hard to imagine how a coherent interconnected body of knowledge will result.

SELF-DIRECTED ACTIVITIES: THINKING ABOUT THE *CURRICULUM STANDARDS*

Standard 1: Mathematics as Problem Solving

The *Standards* state that both the process and the goal of mathematics instruction should be problem solving. In fact, the *Standards* can be interpreted as implying that if an exercise does not involve real problem solving (as opposed to generating answers to routine questions), then it is not mathematics.

1 Come up with a list of criteria that a *problem* must fulfill if it is to be considered a real problem. Then analyze the tasks you present your students. Where do they meet your criteria and where do they fall short? How can you alter, amend, or supplement your curriculum to reflect better this first standard?

Standard 2: Mathematics as Communication

Communication usually is defined as the exchange of thoughts, ideas, and so forth. This implies a reciprocal relationship between the parties involved. One person does not transmit information and the other receive, but both share in the creation of a common meaning. This mutual sharing of ideas also is implied by the word *discourse* as it is used in the *Professional Teaching Standards*.

1 If mathematics is fundamentally a communicative medium—that is, a way in which we share certain types of information—what should a mathematics classroom sound like? What should it look like? What ways other than language can students share what they know?

2 How do you facilitate students' communication of their mathematical understanding in your classroom? How do you hinder it?

3 How do you interact with students who do not speak English as a first language (assuming English is your first language)? How can you enhance their abilities to communicate their understanding of mathematics to you and their peers?

Standard 3: Mathematics as Reasoning

Mrs. Hillman is a fourth-grade teacher. Her class is studying fractions in mathematics, and her students are busy ordering fractions: 1/2 is greater than 1/3, which is greater than 1/4, and so on. Suzanne is using her fraction calculator and hits the wrong button. She shouts, "Mrs. Hillman! I hit the F—>D button and I got .125! When I hit it again, it came up with 1/8. What is it doing?"

1 What questions could you ask of Suzanne that would allow her to reason the meaning of the decimal notation herself?

2 In what ways does technology facilitate students' reasoning in mathematics? In what ways does it hinder reasoning?

3 Examine your interactions with your students. Are you telling them the answers or are you creating questions that allow them to figure it out for themselves? What kinds of questions can you begin asking?

Standard 4: Mathematical Connections

Lesh (1979, cited in Post et al., 1993) created the following model to illustrate how the ways in which we process mathematics are translated from one representation to another as we communicate our ideas (see Figure 3). We can take written symbols (e.g., the equation $y = 2x + 4$) and draw a picture that shows the general structure or pattern underlying the symbolic expression (the graph of a line with a slope of 2 and a y-intercept of 4). We can talk about the picture and describe a real-world situation in which it

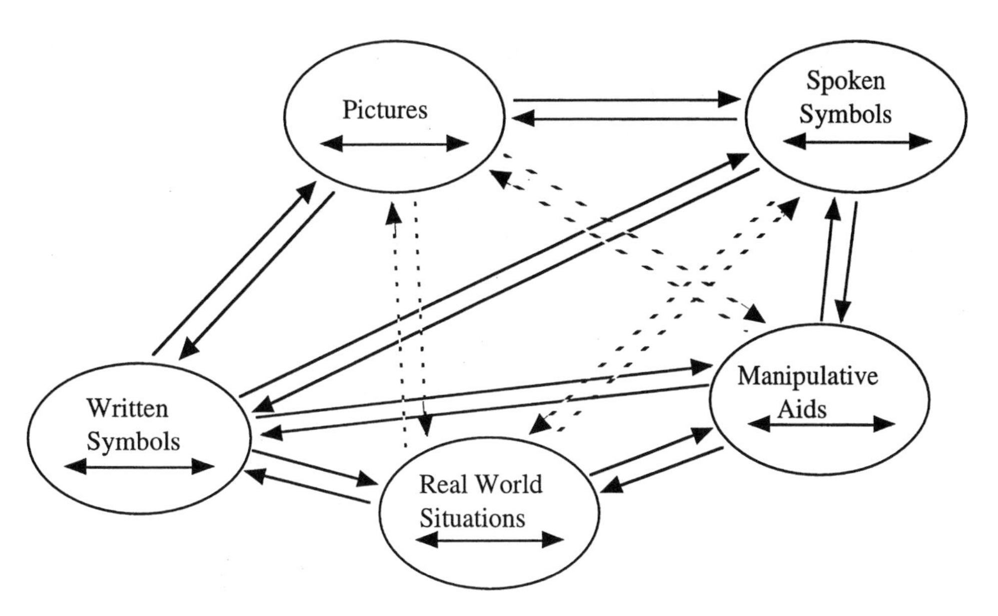

Figure 3 *Lesh's Translation Model of Higher-Order Thinking in Mathematics. Adapted from Post, Cramer, Behr, Lesh, and Harel. In Carpenter, Fennama, and Romberg (Eds.),* Rational Numbers: An Integration of the Research *(1993). Lawrence Erlbaum Associates, Inc., Publishers. Used with permission.*

applies (e.g., "The number of women in a committee on safety must be twice the number of men, and a minimum of four women must be on the committee"). We can grab some jellybeans and put four red ones in a pile and then add one blue and two red to see what the number of "women" (red jellybeans) and the number of "men" (blue jellybeans) will be for each possible committee. Lesh maintained that it is the *translation* from one mode of representing the mathematics to the others that leads to higher-order thinking in mathematics. Explore the following mathematical situation.

> The Problem: A forest management and timber resource company needs to determine if a section of pine trees is ready for harvesting. They have determined that they will only harvest trees with a diameter greater than or equal to 60 cm.

The most common instrument used to measure the diameter of trees is a *diameter tape* (or *D-tape*). The D-tape is a compact pocket tape with a graduated scale on each surface. It is designed with a small hook at the zero end to grab the bark of the tree so a single person can wrap it around the circumference. The D-tape is wrapped around the tree at breast height (about 140 cm above the ground) and the diameter can be read directly from the tape. Trees on a slope are measured on the uphill side.

1 Why do you think trees are usually measured at breast height?

2 Create your own D-tape. What things do you have to take into account to accurately read off the diameter?

3 What if you created an A-tape (a tape that allows you to read off the *area* of the cross section at breast height)? How could you do that?

Students' Reasoning

Group 1 (see Figure 4)
Group 2 (see Figure 5)

1. Well, trees are usually big near the roots
Kind of straight in the middle section,
and skinny at the top. At about 140
cm high, most trees are pretty straight.
So, its kind of like an average or
middle measurement. Not too big, not
too small.

2. We knew that the circumference of a
circle (c) is equal to its diameter
times π. So, we wrote the equation
$c = \pi d$. We then thought, we really want
d, so lets rewrite the equation so d's on
one side and everything else is on the
other side. Since you multiply d by π
to get c, you just divide c by p, and
you know d. Our new equation turns
out to be $d = c \div \pi$. Now, all we do
is for every cm of our tape we
divide the number for the diameter.

3. Using the same strategy, we knew the
area of a circle is equal to π times
the radius squared.

This is $A = \pi r^2$

We also know that $r = d \div 2$.

So, $A = \pi d^2 \div 4$

Since $d = c \div \pi$, $A = \pi (c^2 \div \pi^2) \div 4$

Now on the other side of d tape, we
square each measurement for the
circumference and divide it by 12.56.

Figure 4 *Students' Solution to Problems 1 Through 3 Using Written (Algebraic) Symbols*

Group 2 Using Graphs

1. The roots of trees bulge out at the bottom. The breast height is usually high enough to not bulge as much (unlike one of the members of this group).

2. We knew that $C = \pi d$. But it wasn't so easy to find d neatly, so we made a table:

Circumference (cm)	Diameter (cm)
3.14	1
6.28	2
9.42	3

We saw that for every 1 we add to the diameter, we added 3.14 to the Circumference. We didn't want to keep doing this forever, so we drew a graph:

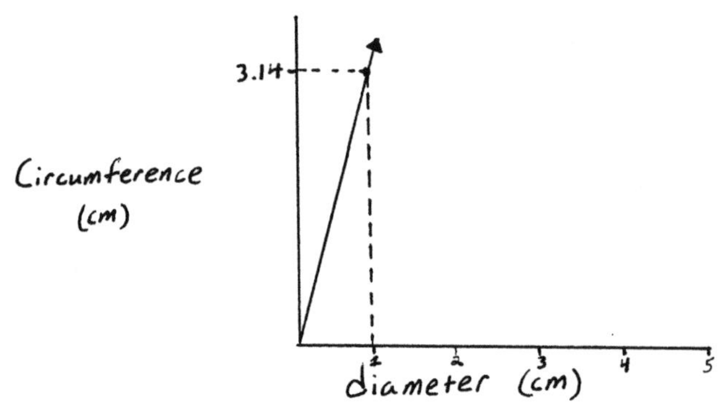

Figure 5 *Students' Solution to Problems 1 Through 3 Using Graphs*

For every cm. in diameter we went over one and up 3.14. We then connected the dots to make the line. Now for every length on our tape (which is really the circumference) we just go to the graph and go over and down to find the diameter. We read it off and mark it on the tape.

3. We did the same thing for the area. We knew that $A = \pi r^2$, so we put in more columns in our table:

Area (cm^2)	Circumference (cm)	Radius (cm)	Diameter (cm)
.79	3.14	.5	1
3.14	6.28	1.0	2
7.07	9.42	1.5	3
12.56	12.56	2.0	4
19.63	15.70	2.5	5

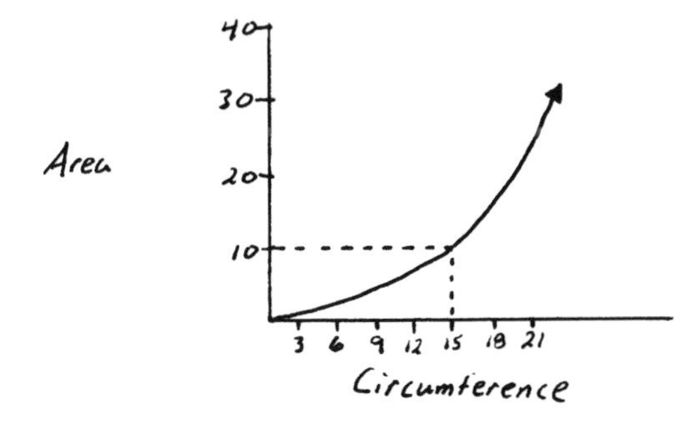

Now we just go over and find our circumference, and then go up and read the Area from the graph and put that on our tape.

Figure 5 *Continued*

TEACHER QUESTIONS

1 Try to express the pattern in each of the modes Lesh presented and describe how they relate to each other. How does this lead to higher-order thinking?

2 How can you help the two groups relate their different modes of thinking to perceive the underlying commonalties?

3 How can you build these types of connections into your curriculum?

4 How are the fields of number, algebra, geometry, and statistics connected to each other? Examine the relationships among these mathematical domains. How can connecting different mathematical domains lead to a fuller understanding of each?

Although teachers determine the actual content of their classes, curricular resources determine to a large degree, what is possible in terms of student activity. Critical examination of the nature of any curriculum should focus both on the ways in which it embodies the historically constructed field of mathematics (the content), and the ways in which it facilitates student inquiry (the process). Two imperatives can guide the evaluation of curricula:

1. Students must have considerable experience building mathematical knowledge informally through the use of models, tools, and discussion prior to moving to more abstract ways of thinking.

2. Multiple ways of representing mathematical ideas must be provided so that students can communicate effectively, and build higher-order understanding of mathematical concepts.

The translation of curricula into meaningful activity is the focus of the next goal: Building Teacher Strategies.

goal three

Building Teaching Strategies

"Maybe we shouldn't be asking, 'Are we doing things right?' but, 'Are we doing the right things?' "

Hedrick Smith, journalist (1995)

In an address to a National Press Club Luncheon on July 12, 1995, Hedrick Smith, noted author, journalist, and political commentator described the American Educational System as being irrelevant to the future life of most students. His research revealed that students are learning with outmoded tools and techniques that do not give them appropriate experiences to make a smooth transition into the workplace. Smith emphasizes that we are currently teaching our children pretty well. The problem is, we are teaching them the wrong kinds of things.

Goal One discussed the nature of *what* should be taught. Goal Two will deal with *how* a good teacher can take the right kinds of things and transform them into good teaching.

THINKING ABOUT TEACHING PHILOSOPHIES

In addition to selecting curricular resources, implementing reform in an intelligent way requires a different approach to the things we do as teachers. Table 2 illustrates some of the conceptual shifts that facilitate implementation of the *Standards*.

Prior to teaching contextualized, meaningful mathematics, it is extremely important that teachers do their homework. It takes time, effort, and especially *attitude*. Change is not easy, and it will take perseverance. We do not say this to frighten anyone from changing. Rather, we intend to alert you to prepare to take the vital steps that will make this a successful mission.

Part of your preplanning activity should include developing the background of the philosophy of teaching to which you will ascribe. The *Standards* are a good place to start, but they are too general to translate directly to all styles of teaching and all levels of knowledge and to maintain a comfort zone for all teachers. Try to answer the following questions *for yourself:*

1 How does this new philosophy differ from traditional mathematics textbook teaching?

Table 2 Conceptual Shifts in Teaching for Understanding

Away From	To
Covering the material	Facilitating meaningful activity
Mastery as the goal of instruction	The search for understanding as the goal of instruction
The teacher's way of knowing as the valued way	The students' ways of knowing as the valued ways, supported by logic and verification
Classrooms as collections of isolated individuals	Classrooms as communities of learners working together
Authority by edict	Authority by reason
Quizzing as the primary method of assessment	Questioning as the primary method of assessment
Lecture and regurgitation	Discourse

2 Why is it better for your students? Why is it better for you as their teacher and fellow learner?

3 What exactly is the vision? Figure out where you are in relation to the vision and devise a plan that will allow you to begin to attain that vision.

We did not just jump into teaching. We were trained for years in college, and then tried our hands at it with some supervision. In addition, those of use who have been out in the schools for some time have developed routines for classroom management and teaching content with which we feel comfortable. It took years to develop these attitudes, knowledge, and routines. Do not expect to overturn all of this training and your own personal experiences at

BOX 1 **Preplanning Activities for Dealing With Change**

1. Understand what you want your students to learn. Explore avenues that will help you facilitate that learning.
2. Read all you can about the *Standards* and the teaching that is envisioned in these documents.
3. Talk with fellow teachers who are involved in this type of teaching already. (Pick their brains.)
4. Attend some conferences or seminars, preferably with *hands-on* workshops. (Remember, do the math like a student yourself.)
5. Find a 1- to 2-day activity that will fit into your curriculum. Try the activity and reflect on your successes (and failures).
6. Find a mini unit or more activities that fit together and lead somewhere mathematically. (Notice the shift to a set of coherent experiences.)
7. Try a curriculum unit (2–4 weeks of activity that fit together mathematically and thematically). Later, try another.
8. Hook up with someone—by telephone, computer modem, or face to face—who has been teaching in this way for a few years for professional support. Other teachers are your best resource.

once. Pace yourself. Do not set yourself up for failure by neglecting to prepare for this change. Test the waters and then jump in with some knowledge of the territory. The preplanning activities presented in Box 1 have been helpful for us and our colleagues who are struggling with this change process.

Eventually, you will mold most—if not all—of your teaching toward the vision only to find that the vision has changed. To keep dynamic, you must begin to see change as an ongoing process that never ends in our quest to become better teachers.

Doing mathematics should go hand in hand with preparation and planning for math class. To understand the problems your students will run into, it is vital that *you* do the work prior to presenting them with problems to solve. This may seem obvious, but it is fundamental to making appropriate plans: anticipating student responses, creating probing questions, and projecting possible ways in which the lesson will progress. We probably have learned more and understood more about the realm of mathematics since we began planning this way than in all of our years of education culminating to that point. Those of us who were "good" at mathematics probably found math fairly easy. We could memorize the algorithm and do the problems in the book or worksheet. We seldom had to apply the skill or create methods for problem solving ourselves. Now, however, we are asking our students *to think mathematically*, something we were probably not taught to do. It is important we go through this process as well.

After becoming familiar with the content and process of the unit, it is time to gather all of the tools you will need. Any papers that need to be duplicated, manipulatives needed, supplies that need to be on hand, should be there for you to use as you engage in problem solving.

PLANNING

Research indicates that teachers engage in eight different levels of plans as they map out their course of instruction for a year. These planning levels include daily, lesson, weekly, unit, short-range, long-range, term, and yearly plans (Clark & Peterson, 1986). Although all levels of planning are important for teachers, we will focus on the three most important types here: unit, weekly, and daily.

One thing to keep in mind as you begin planning your reform-oriented lessons is that most plans are not official plans put on paper but rather are *goals* teachers have for themselves and their students. Most plans are done mentally, and in fact, most are mere outlines or lists of topics instead of specific objectives to be covered or mastered. Coherent written plans are developed primarily to appease administration, to ensure that there is some flow of ideas when a substitute teacher is present, and to remind you next year of what you did and how it went. This is not to say that you should not write down your plans. You should. However, you should develop your own plans, in your own style, that may or may not translate easily to the principal.

The functions of plans are to prepare the *teacher* cognitively and instrumentally for instruction. The mental image of the lesson plan is used to guide the teacher's behavior during routine instruction (Clark & Peterson, 1986). In essence, if you plan well, you will not have to keep all of the details of a lesson in mind all of the time—your plan will allow you to focus on the bigger issues, on the immediate dialogue you are having with your students—and it will help you guide the interactive process of instruction.

One of the most important things to keep in mind as you actually begin teaching a planned lesson is that nothing ever goes exactly as planned. Students will get hung up on ideas you think trivial, and they will understand concepts you can hardly imagine. Abandon your lesson plans when activity flow is threatened and use what you know about your students' prior knowledge and your larger mathematical goals to guide you.

Teachers depend heavily on published teachers' guides or textbook structures to help them make certain that the mathematics they are teaching fits into a coherent, developmental framework. This reliance has its benefits and its problems. On the positive side, having an activity or lesson structure outlined for the entire term enables the teacher to focus on the day-to-day process of teaching, comfortable that the larger picture has been addressed. The authors of the *Standards* do not expect teachers to develop all of their activities, their vision of mathematics, and their assessment practices alone without material help. Textbooks and other materials help fill this role.

On the negative side, we have historically been overreliant on textbooks to dictate the mathematics we teach. Take a look at a typical textbook for the grade you teach. Most books contain 130 to 150 lessons. Note that given the 180-day school year, this means one lesson per day. Also note that because most textbooks are about 300 pages long, a typical lesson is usually displayed on a two-page spread. Very few lessons can be of the extended type.

Where is there room for going in depth on larger, more interesting mathematical issues? These are just physical attributes of most textbook series. What about the content? Think back to goal two and discuss with a colleague how well your current materials fit the vision of the *Standards*.

Box 2 lists a nine-step process for developing lesson plans that stems from the research findings summarized in this book. Although these processes have been presented as a nine-step procedure, the reality of planning should not be so rigid. Often the prior knowledge of the students will determine the types of activities you choose. Many times the classroom environment will put constraints on what you are able to develop. In such cases, these considerations may take precedence over the other considerations. Treat these processes as questions to keep in mind as you go about planning your lessons.

A note about assessment (step number 8): It is important that your assessment techniques be tied explicitly to the objectives of the lesson, and in turn, to the *Standards* that helped you define your objectives. For each objective, ask yourself the following questions (taken from the *Assessment Standards*, NCTM, 1995, p. 14):

- How does the assessment contribute to each student's learning of mathematics? Examine the description of the important mathematics in the *Curriculum Standards* for the topic you are teaching.

- How does the assessment relate to instruction? The assessment should reflect the content, processes, and contexts you use in your teaching.

BOX 2 **A Nine-Step Process for Developing Lesson Plans**

1. Pick a topic.
   - This should be done carefully, so that all your lessons interconnect meaningfully over the year.
   - Pay specific attention to the mathematical coherence of the topic within the yearly sequence.
2. Determine the prior knowledge of the students.
   - What have they done before mathematically?
   - What have they done before experientially?
   - What modes of thinking are used by different students?
3. State the goals clearly.
   - Introduce a new topic.
   - Develop understanding.
   - Review old knowledge.
4. Determine the instructional sequence.
   - What ideas logically flow from the prior knowledge of the students?
   - How can you incorporate experiential models (such as a number line made of beads, a fraction bar, etc.) so that all students can have immediate access to the mathematics?
   - How do you project students' thinking will change as a result of activity with the models?
5. Decide on the methods of classroom organization.
   - Will the whole class work together?

Continued

- Will the students branch off in cooperative groups?
- Will they work as individuals?

6. Map out the procedures to be followed.
   - How should the topic be introduced?
   - Will you have to review any prior material?
   - What probing questions will you be prepared to ask as students struggle with the material?
   - What materials (text, units, etc.) will be used?
   - Prepare a couple of different options for homework you will use, depending on how students approach the topic.
7. Decide how much time to spend on each part of the lesson.
   - Be flexible; your first year, you will find students take longer than you expect.
8. Decide on the assessment techniques you will use.
   - Will you use probing questions?
   - What will you focus on in your observations?
   - What tasks will you set up specifically to see what children know?
   - How will the assessment further the students' learning?
   - How will it help you teach better?
9. Write down the lesson plan.
   - Do not be too specific. Allow for flexibility.
   - Make sure it is understandable by others' with a vested interest in education, including other teachers, the administration, and parents.

- How does the assessment allow students to demonstrate what they know and what they can do in novel situations? Provide avenues for students to go beyond what you explicitly did in class by using different situations and allowing them to project to more complex mathematical concepts.

- How does the assessment engage students in relevant, purposeful work on worthwhile mathematical activities? The problems you create for assessment should reflect the model of problem solving, reasoning, communication, and connections that underlies all worthwhile mathematical activity. It also should motivate students to do their best by using relevant contexts and challenging content.

- How does the assessment build on each student's understanding, interests, and experiences? You must design tasks in which all students have access to success through the use of multiple models, ways of thinking and interpreting information, and displaying their solutions.

- How does the assessment involve students in selecting activities, applying performance criteria, and using results? Create a range of activities that students can choose from to show you what they know. Make sure that they can articulate what good work looks like and how it can be communicated to others.

- How does the assessment provide opportunities for students to evaluate, reflect on, and improve their own work—that is, to become independent learners? The test or other assessment activity you design should not end when students turn in their papers. Use their work to help them understand the areas they need to improve.

If you can answer all of these questions and design your assessment to address the content in the *Curriculum Standards*, your lesson plans will flow naturally through the introduction, development, and evaluation cycle.

Notice that we have placed writing down your plans last on the list. Written plans are by nature telegraphic—that is, they do not convey adequately all of the real planning you have done. Thus, written plans function primarily to orient you to the larger understanding of the activity you are envisioning and to enable others (such as substitute teachers) to get a handle on what the children are experiencing.

QUESTIONING

As teachers, we have been trained to have all of the answers (and the *right* ones). In the past, we have prided ourselves in knowing the right answer and being able to tell our students immediately the answer when they became stumped. We rarely focused on the process of problem solving or the strategies students develop and use to solve important problems. Those students who were good memorizers were the students who soared on the tests and homework. However, this did not mean they understood the concept or they knew how to apply the skills in question.

Today, we need to hold back our answers and provide our students with more questions that will lead *them* to find an appropriate solution. This encourages children and gives them more mathematical power, which in turn enlightens their way rather than merely taking them to the end without thought of the journey. When interacting with our students, the two simple words *how* and *why* open up the window to a vast understanding of what our students actually know. Having to explain how they arrived at their answers or why they did something that way helps us as teachers to see where they are in their thinking and understanding of the concepts and skills.

A student of ours—Shelly—had little to no self-confidence in the area of mathematics. She was typically described as a C or D student; mathematics had not been easy for her, and she struggled to attain a C grade. One of our projects for the concept of area was to figure out the area of different parts of the classroom. As Polly was walking around to the different groups, she asked Shelly how she figured out the area of one ceiling tile. (It is important for you to know that my classroom is made of cement block walls (40 cm × 20 cm each), square 10-cm floor tiles, and ceiling tiles that are about 61 cm × 122 cm.) Shelly's group had already measured the cement blocks that make up the walls and had estimated the area of the wall. A couple of the members of her group were standing on the desks trying to measure the ceiling tile. Shelly indicated that she had an alternate method. I asked her to explain how she would go about solving the problem:

> If you look at the ceiling tile and line it up with the cement blocks on the wall, you can see that it equals 3 lengths of the blocks. And if you look at the other wall, you can see that the width of the ceiling tile is equal to one and a half lengths of the blocks.

I had never thought of doing the problem that way. So we sat down and I asked Shelly, "How long is a block?"

> Shelly: 40 cm.
>
> Teacher: So, how long would the ceiling tile be?
>
> Shelly: 80 cm . . . no . . . 3 times would be 120 cm. (Note the actual length was about 122 cm; hers was a pretty good estimate.)
>
> Teacher: How wide would the ceiling tile be?
>
> Shelly: 40 cm plus 1/2 would be 80 . . . no 60 cm. (Note the actual width was about 61 cm; another good estimate.)

Her reasoning indicated that Shelly really had thought about this problem and was applying what she already knew to help her with the ceiling portion of the problem. Her way of doing this problem was different than anyone else's, and was a creative idea; all of the other groups were measuring each of the tiles manually (which was not wrong—it was also a good strategy). Shelly, who shined very rarely in math, glowed this day and gained a bit of self-confidence with regard to her math ability. These little "ah-has" spark us as teachers to interact with our students more and to ask those questions: *how* and *why*.

SELF-DIRECTED ACTIVITIES: THINKING ABOUT THE *TEACHING STANDARDS*

Standard 1: Worthwhile Mathematical Tasks

When teachers examine the materials they use compared to the *Standards*, they often perceive that the slick packaging and colorful graphics are window dressing. Looking at what children are actually asked to do and what is expected of them, they see that most—if not all—textbooks fall far short of the *Standards* in both content and philosophy. For example, if you only look at the answers in a teacher's guide, what do you see? Red numbers? If you only see red numbers or one-word answers, what does that tell you about what the textbook is trying to teach? What about process? What about diverse strategies? What about *communication*?

1. Examine the textbook series or curriculum materials adopted by teachers in your school. Contrast the materials with the *Curriculum and Evaluation Standards*. Where do they fall short? What can you do to remedy this in your own classroom?

Standards 2 and 3: The Teacher's Role in Discourse and the Students' Role in Discourse

The *Teaching Standards* state

> Because the discourse of the mathematics class reflects messages about what it means to know mathematics, what makes something true or reasonable, and what doing mathematics entails, it is central to both *what* students learn about mathematics as well as *how* they learn it. Therefore, the discourse of the mathematics class should be founded on mathematical ways of knowing and ways of communicating. (NCTM, 1991, p. 54)

In the first author's beginning elementary mathematics methods course, my students become engaged in worthwhile problems—problems that are connected to important real-world situations and that are connected mathematically. They experience good, challenging problems from a student's perspective. In doing so, they often become so engrossed in the activities that they fail to extend their learning to the actual methods part. They lament, "How do I begin to develop a repertoire of strategies that will enable *my* future students to experience mathematics this way?" They often find it difficult to ask questions of their students and instead resort to telling the students what to do. This gets the students on task but takes the problem solving, the reasoning, and ultimately the mathematical power away from their kids.

To help students on their way, we devote a large portion of time to developing generic question formats that can be tailored to the tasks students are facing. For example, one such generic format is, "I see that you did ___________. How did you come up with that?" This question reflects the students' strategies back to them, and then requires them to explain the reasoning behind the strategy. A second generic question they have found useful is, "Ah, so you did ___________. What if I did this?" (Then change the situation slightly so the student has to draw distinctions between what they are currently thinking and where you want them to go.)

Consider the following vignette: In a third-grade classroom, the teacher, Mr. Sanchez, poses the following problem on equivalent fractions:

> "I want you to draw me a picture of two tables. The first table has 8 chairs and 4 people are sitting there. The second table has 10 chairs and 5 people are sitting there. Which table is more full?"
>
> Anne: "The second table is more full."
> Jodi: "No! There are more places there!"

1 What questions could you come up with to help Anne and Jodi resolve their conflict?

2 Think up at least 10 generic format questions that you feel will be helpful for you in orchestrating meaningful discourse in your classroom. As you develop your lesson plans, fill in the blanks. Anticipate the misconceptions your students may have, the strategies you expect to see, and the types of reasoning you feel your students will engage in. Create probes you can use to keep the dialogue moving.

Standard 4: Tools for Enhancing Discourse

The notion of a tool as suggested by the *Teaching Standards* is much broader than mere technologies (see Pea, 1987). The authors of the *Standards* suggest that stories, models, modes of presentation, and symbols as well as calculators and computers are all viable tools the teacher of mathematics must use at appropriate times in teaching. In its most basic form, the definition of a tool can be anything you can use to make kids' problem solving, reasoning, communication, and connection of mathematical ideas more fruitful (notice we did not say *easier*).

1

List the technological tools you feel would be helpful for learning mathematical content. Include different types of calculators, computers, other peripherals for computing technology, and software. Also include manipulatives and hands-on tools.

2

Now focus on those tools you do not currently have available (or that are currently impractical to use because they are not in your classroom). Prioritize these tools, and come up with a plan to get them. This may entail writing to computer or software companies to get demonstration copies for you to review. It may also entail purchasing a personal version so that you can become familiar with their use. In creating your plan, remember to develop a sales package that effectively allows you to make your case to the budgeting personnel in your district.

Standard 5: The Learning Environment

The classroom environment as a whole can be thought of as consisting of three key components: (a) the physical environment, (b) the intellectual environment, and (c) the social environment. For the physical environment, the types of seats students sit in can make a huge difference in the ways in which the teacher can organize group activities.

1. Contrast the benefits of having a classroom with 36 individual desks with attached chairs, 36 individual desks with chairs that are not attached, or a set of 9 tables with chairs that are not attached in facilitating different physical learning environments. (Hint: Draw a picture!)

2. The *Standards* emphasize that a variety of tools be available for students to use in solving problems. These tools include (but are not limited to) computers, calculators, graph paper, counters, scissors, glue, tape, and so on. Whatever a student can use to make sense of a problem should be available to them. What impact does this have on the physical environment of the classroom? How can you organize your tools to be more available to students, yet not disappear on you?

The intellectual environment of a good mathematics classroom involves engaging students in intellectual risk-taking, conjecture and verification, and building relational knowledge. "Serious mathematical thinking takes time as well as intellectual courage and skills" (NCTM, 1991, p. 58).

3. Given that most teachers have only about 45 minutes per day to spend on separate subjects, how can you better manage your time to set up this type of intellectual environment? (What connections can you make, both content-wise, contextually, and with other human resources?)

The social environment of your class will have serious consequences on students' learning and, perhaps even more importantly, on their feelings and dispositions to continue in mathematical activity.

4 What social rules can you develop with your students to foster intellectual risk taking in a psychologically safe environment?

5 What are the roles of the teacher and the student within such a classroom?

Helping students learn sound and significant mathematics requires change in three fundamental aspects of our practice: planning, teaching, and assessment. Our plans must change to reflect better the prior knowledge of our students, how we can effectively manage the tools for enhancing discourse, and how the mathematics is connected to other subjects. Our practice must change to include more social learning, necessitating a shift from teaching-as-telling to teaching-as-questioning. The questions we ask them can form the basis for our assessment strategy. We will examine this shift in assessment practice in Goal Four.

goal four

Building a Balanced Assessment Strategy

The Heisenberg uncertainty principle of physics states that we can only know either the position of an electron in space, or its path, but not both simultaneously. Compounding this difficulty is the fact that by studying the phenomenon, we fundamentally alter its nature: The perspective of the observer cannot be separated from the phenomenon being observed. This uncertainty is also present in the assessment of students' knowledge. The teacher cannot know everything about a student, and by asking a question to assess what a student understands, the teacher alters the very thing she is trying to figure out! This Goal attempts to unravel some of this uncertainty. Instead of thinking of knowledge as a position in time, we encourage you to think of knowledge as a trajectory: always moving forward.

> As a diagnostician, the teacher is trying to map into his own the momentary state and trajectory of another mind and then, as provisioner, to enhance (not replace) the resources of that mind from his own store of knowledge and skill. (Hawkins, 1972, p. 112, cited in Ball, 1993)

Assessing and evaluating students is always a concern for teachers. The greatest difference from a teacher's perspective that occurs with implementation of the *Assessment Standards* is that assessing students' progress is continuous; it is not based on a unit test at the end of each chapter. More and more of the evaluation is built around the participation and involvement of the student. The evaluation process has fanned out to incorporate more than answers to problems that are either right or wrong but also reflects the first four standards enumerated in *Curriculum Standards* problem solving, communication, reasoning, and connections.

In managing this type of assessment the teacher must allow the time to have discussions and give students the opportunity to express how they came to their conclusions. Many times, given the chance to look at a problem differently, even confused students can discover that they are on the right track mathematically. Here is an example: In a unit we are currently working on called *Ways to Go* (de Lange, de Jong, & Spence, in press), we are comparing distances and times to get from one city to another on a map. The question asked is, "Which city is farther from Redding, Eugene or Bend?"

From the map, students need to determine the following:

> From Redding, California, to Eugene, Oregon, the distance is 315 miles. The time it takes to make the trip is approximately 5 hours and 13 minutes. From Redding to Bend, Oregon, the distance is 281 miles and the trip takes about 5 hours and 33 minutes.

Depending on the students' interpretation of the word *farther*, either town would be counted as correct, *if* the students had a logical reason for that answer based on their mathematical problem solving. Real understanding of the situation is shown when the students conclude, "It would depend if you were short on time or if your car was really old with a lot of miles or if you were short on gas." This indicates that the students perceived the physical distance and the elapsed time as both being valid metrics for essentially a measurement and rate problem. With questions like this, dialogue develops in the classroom, and students defend their answers with mathematical data. This is what we are striving for: the reasoned use of math. It is important to make note of students' reasoning as you roam the classroom listening to the dialogue and interjecting questions to stimulate discussion.

A PERSONAL NOTE FROM POLLY: CHOOSING A STRATEGY

From a classroom management perspective, I have used many trial-and-error methods in determining the best way to account for a student's progress. To this point, I have chosen to keep track of students' understanding in this way: I have a clipboard with each of the students' names, and I record daily a "+" for those students who have done their homework and a "0" for those who come in with empty sheets. As we go over the assignment, students are to correct their work (if they have done it) or enter their solution strategies on their blank sheets. They may turn in this work for minimal credit. At the end of the quarter, they are given a grade for the percentage of daily work completed on time. During group work time, students are given participation points for staying on task and working in *their* group (some students enjoy working in others' groups). These are then totaled for a participation grade.

Usually, on Friday I give a quiz that covers the material we have discussed Monday through Thursday. Some of the questions come directly from the homework; some are almost the same, possibly with just the numbers or figures changed; others ask them to apply what they should know to new situations. At quiz time, the students are allowed to use their homework, curriculum units, and calculators. If they are paying attention during class and doing their homework, and if I have done a good job of stimulating their thinking, they should be able to work through the quiz without difficulty.

Another section in my grade book is for graded homework. This section includes any problems from the packet we are working on that I feel embody the important concepts. Any problems we have used for skill reminders or brush-up also go in this section.

New to my list this past quarter are basic knowledge tests. I give out some refresher worksheets on Monday and the students have a quiz on the material on Friday. These tests review basic metric measures, benchmark fractions and their corresponding percentages, and other pieces of information important for people to know and use in an information age. The reason I include such basic items in my assessment strategy is that many times when we are working in a unit, my students are stumped because they do not know how many days are in a year or how many millimeters in a meter. These stumbling blocks are avoided by some refreshment prior to the time when they are needed.

I use all of these pieces of information to help me determine what students know and can do. The graded homework gives me information on the students' conceptual understanding of the topic at hand. The basic knowledge quizzes help me determine where students may be faltering in their problem solving because they may have forgotten a few simple facts or relationships. The group participation points help students work more diligently and help me determine who is doing the work and who should be grouped together for the next activity. They all add to my ability to teach the students better.

This strategy is working for me. However, I am sure it will change with time. The key question I am struggling with is, "Will the strategy I have chosen give the student and parents a true reflection of the student's progress?"

ASSESSMENT VERSUS EVALUATION

When the authors of this book first began implementing a pilot curriculum aimed at meeting the *Standards*, we were discussing the nature of students' understanding of measurement and pre-algebra concepts. Polly reflected, "I know these kids have learned more in the past 3 weeks than they have in the whole semester, but I don't know how to assess it!" This sentiment is a common frustration teachers experience when they begin to value different nonstandard strategies, argument and conjecture as evidence of knowledge, and the use of manipulatives and other ways of modeling mathematical ideas.

To get a handle on the nature of assessment and evaluation in a reform-oriented classroom, it is helpful to distinguish between the two. *Assessment* refers to the "comprehensive accounting of an individual's or group's functioning within mathematics or in the application of mathematics" (Webb, 1992, pp. 662–663). This definition, when coupled with the philosophical orientation of the *Standards*, implies studying students' performance in a wide variety of contexts, using a wide variety of methods. Teachers should attempt to use whatever means are at their disposal to get at what students know and what they can do. These methods include qualitative approaches such as observation and interview (looking and listening), as well as quantitative approaches such as quizzes and tests (assessment events) (Webb, 1992).

Evaluation refers to the assignment of value to students' performances (Webb, 1992). Included under this definition are methods for grading, determining accountability, and academic placement. Assessment, then, *can* serve as a medium for evaluation, but it is a broader, more inclusive domain that ultimately should relate directly to informing instruction.

At the most basic level, we gain most of our information about students' learning through our informal interaction with them. We ask questions and listen to students' answers. These responses are then used to frame other questions, which in turn lead to students' thinking about the task in different ways. The questions we ask are often found in the form of curricular tasks, but just as often they are created spontaneously to make the most of a teachable moment. The following excerpt is from a beginning first-grade teacher's daily log. It is a good example of how a teacher has struggled with finding out what her students know and where that should lead instructionally. As you read through it, try to place yourself in the shoes of the teacher. What are the issues she is struggling with? What information does she attend to? How does she use her reflections to frame her teaching?

A FIRST-GRADE TEACHER'S REFLECTIONS

September 16

I wasn't sure where to begin and I made a lot of mistakes. In fact, I'm not sure if I did too much that was right.

I hadn't really worked with any of the students in problem solving or mathematics, only in reading. I used the information from my work with the students in reading to try to help me make a decision about the four students to pick for my group[s]. I tried to make the groups as heterogeneous as possible. I also tried to choose students that I thought might be at various levels. Beth and Randall are very bright and personable children. Both are good readers and get along well with the other students. I thought that they might lead the group.

Magdalena is very quiet and shy. She struggles with reading and has difficulty voicing her opinion. I am hoping the group might draw her out. Salvino doesn't talk. He struggles with his class work and doesn't read much yet. I believe he might really have to struggle with math, and I am hoping that the group will help him.

I hope that this group would work well together, but I'm not so sure that it will. The kids were talking amongst themselves, but not about mathematics, and I had a hard time getting them to focus on what I was trying to do. It could be that I didn't really know what I was doing, so the students were confused too.

I collected some manipulatives—kidney beans and unifix cubes—and I had them on hand. I had initially planned to start out with some simple problems and try to acclimate the students to explaining their strategies for solving them. I knew that they probably wouldn't have any experience with explaining their thinking and that it would be difficult for them. I had no idea how difficult. Three out of the four kids easily answered my questions but they could not explain to me how they got the answer—only that they got them. I tried to drop the difficulty of the task way down to make it easier for them to explain their methods to me, but they couldn't. Randall, who seems to calculate simple addition and subtraction (10 or less) rather quickly in his head had a very difficult time trying to tell me how he got the answer. I kept asking probing questions—sometimes leading questions—and at one point, Randall made up something that didn't make sense to explain an answer he had gotten. I think he just felt that he had to say something so he made something up (his explanation had very little to do with the problem he solved).

I decided to switch my strategy. I had been giving them all addition and subtraction problems, and all but Salvino (who just sat there smiling at me and refused to participate) were able to give me an answer seemingly without even counting on their fingers. I decided to try what they call a hand assessment. I had five kidney beans and showed them to Magdalena. She counted them and then I pulled them back and hid two in one hand and showed her the hand with three. I asked her how many I was hiding. Without even batting an eyelash, she told me two. I asked her how she knew that and she just shrugged her shoulders and smiled at me shyly. None of my further probing could get her to admit how she got that answer. She would only say that she "just knew." I had the same results with the other students as well. For the most part, they could give me the correct answer but they couldn't explain how. They just looked at me with an expression that I interpreted as, "What do you want to hear?" Not only could they not explain their methods to me, I could not see any indication of how they were doing it either. I didn't see them count the objects or use their fingers. I didn't know what to do.

I decided to use larger numbers and I put ten beans in my hand. I hid eight and showed two to Randall. I asked him how many were in the other hand. He answered eight immediately. When I asked him to show me out loud how he thought in his head, I finally got him to explain his thinking to me. He said that he knew there was ten and that he counted the beans in my hand backwards—ten, nine—so there must be eight left in the other hand. I was ecstatic that I finally got an answer and I probably went overboard with the praise. The other children simply tried to explain their strategies the same way after that instead of telling me their own strategies.

Frustrated again, I decided to make the problems harder so that I could hopefully see the strategies being used. This didn't work either, however. The students often answered three-step problems quicker than I expected and without any outward sign of how they were getting them. I still couldn't get them to explain how they were getting the answer either. I was really frustrated. I didn't know if they didn't know how to explain what they were doing or if they were afraid I wanted a right answer and were afraid they weren't doing it right.

At the end of 25 minutes, I didn't feel I had accomplished much of anything. The group had talked too much—but not about problem solving. I did discover some of the students' capabilities in problem solving but I couldn't get them to verbalize their reasoning (and I couldn't see it either).

I think for next week, I need to see if I can find anything in the book that might help me create better problems—ones that will be better for the children to solve as a group and ones that encourage a variety of methods. I also want to tape-record the session so that I can have an honest account of my questions and reactions. I don't think I say the appropriate things or ask the best questions and it is too easy for me to be easier with myself when simply remembering my words from memory. I think if I have it on tape, I can analyze myself better and be more honestly critical.

September 23

My objective for this session was to get the students working together and to finally get them to explain their thinking aloud. I decided to do this using *less* and *more*. I asked the group about the meaning of those two words and we talked about two piles of unifix cubes that I had in front of me. One was obviously a bigger pile than the other and they told me easily, without counting them, which had more and which had less. They could verbalize that the one that was bigger had more and the one that was smaller had less. I then had them work in pairs (it happened that the two girls were sitting together so the two girls worked together and the two boys worked together). I gave them similar piles of cubes and asked them to work together to figure out who had more and who had less. Beth and Magdalena each counted their own, talked to each other about how many they had and decided who had more. When I asked them how they figured it out, Beth explained to me that they each counted their own and that they both knew that 15 was more than 13, so that Magdalena had more. I asked Randall and Salvino who had more and Randall told me that he did. When I asked him how they knew, he told me that he had counted them. Unlike Beth and Magdalena who had worked together, Randall had decided to count both his pile and Salvino's. I told Randall that it might be easier if he let Salvino help him, and that they work as a team. They had watched Beth and Magdalena and when I gave them each another set of cubes, they each counted their own, said the number out loud, and Randall announced that Salvino had more. I asked Salvino if he agreed with that answer and he nodded yes. I then told the students that this time I wanted them to come up with another way besides simply counting the cubes to determine who had more. I gave each individual another pile of unifix cubes and asked them to find a way of arranging the cubes to see if they could determine who had the most. I knew that they had done a worksheet on matching to determine more and fewer and I thought they might do that. Instead, they surprised me.

Beth and Magdalena lined their cubes up on either side of one of their pencils. One line was a little bit longer than the other one (see Figure 6). When I looked over toward them I asked them to explain what they did. Beth did most of the talking. She told me that they lined Magdalena's cubes on one side of the pencil and hers on the other. She said that her line was longer and so she had more. I was ecstatic! I finally had an explanation for a strategy and an original strategy at that. I told the girls that was great and I gave them each a different pile of cubes. I asked them

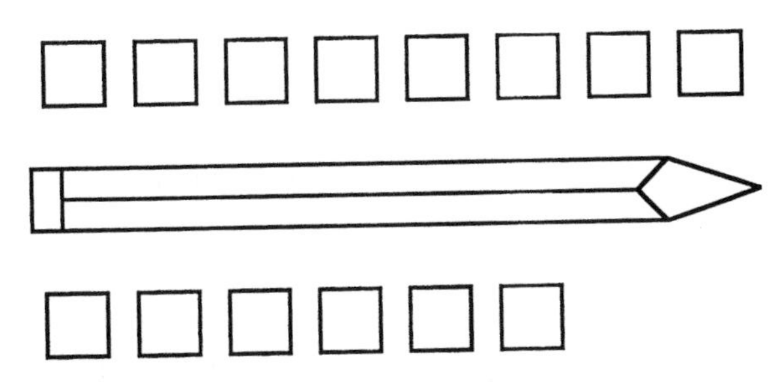

Figure 6 *Beth and Magdalena's Strategy*

to see if they could come up with yet another way to determine who had more.

I watched them discuss it and play around with the cubes for a minute and then I turned back to Randall and Salvino. They were having a difficult time figuring out without counting. I asked them who had more and Randall told me he did. When I asked him why they thought that he explained that he had two piles of five which made ten and that Salvino had two piles of four which only made eight. He had still counted (and Randall had still done all the work), but he had grouped them differently. Beth and Magdalena were really struggling at this point to come up with another way to figure out who had more. I had Randall explain how he and Salvino did it, and unlike the last time, both Magdalena and Beth listened patiently for Randall to explain himself.

I asked them if they could maybe match up the groups to see who had more. They all looked at me puzzled so I showed them what I meant with a few of Randall's and Salvino's cubes. I asked Salvino to finish matching them up for me and then decide who had more. He did and then told me that he knew he had more because his tower was longer. This matching example seemed to set them off. When I asked if they could come up with even more ways, I think they started to view it as a game. I had to keep encouraging Randall and Salvino to work together and help each other instead of Randall doing all the work, but both of the groups kept coming up with more ways to arrange the cubes to figure out who had more. They arranged them in twos and threes and they created something they called the "brick wall"

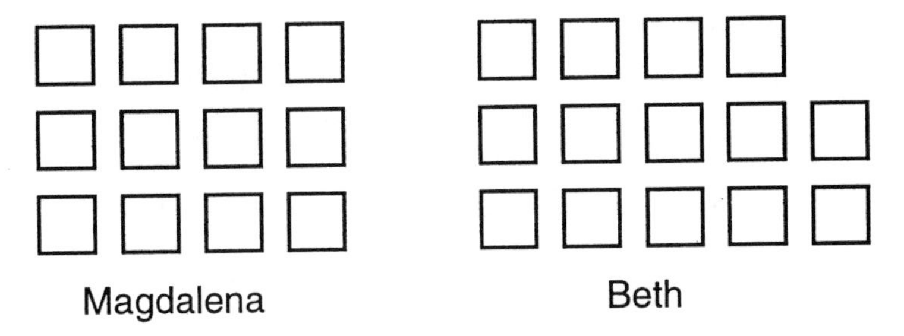

Figure 7 *The "Brick-Wall" Strategy*

which they seemed to like the best. (See Figure 7.)

I started asking the groups how many more one person had than another, and they explained that they counted the cubes that didn't match up. I was thrilled. Not only were they working together to get an answer, they were having fun, they were trying different strategies, and they were explaining their answers.

I think the problems (more and less) were better at encouraging the students to work together then what I tried last week. The students seemed more interested in what they were doing, and their conversations were exclusively about figuring out the problems. They were able to explain their different methods for finding a solution, and they seemed to think it was a game to try to come up with different ways to solve the problems. I think this really helped them to understand that there are various ways to solve similar problems and that they are all good. The students also began to understand that I wasn't looking for just an answer and they learned to explain their methods to me. This habit of explanation has carried over into their reading as well.

I still had some problems, however. I wonder if I hadn't demonstrated the matching and would have instead asked better leading questions, if the students would have eventually come up with that strategy on their own. I'm not always sure when to tell them how they might do something and when to let them struggle longer. As I listen to myself on the tape, I find that I often step in and show the students where they made a mistake instead of questioning them until they discover it themselves. Especially with Salvino, who struggles with most of the problems and who lets Randall figure them out for him, I find myself stepping in and showing him how to do it.

For the next lesson, I want to work on asking better questions. I want to try to let the students do more of the explanations to one another and I want them to experiment a little more with strategies.

September 30

I didn't know where to go from last week's lesson. I decided to move from less and more—and the matching they did to figure that out—to graphing. I wanted the kids to be able to look at a graph and be able to read and determine what it was saying.

I started the lesson off with a review of what we did the time before. I changed the group and put Beth with Salvino and Randall with Magdalena. I gave each student a pile of unifix cubes and asked the two groups to determine who had more and who had less. After they determined who had more using their own methods, I asked them if they could show me who had more by matching them up. Randall and Magdalena did this rather easily, but Salvino didn't understand and he instead just moved his and Beth's cubes around on the table mixing them all up and not being able to figure out whose was whose (I guess I should have had one color for each person in order to avoid this problem). I gave them each a different bunch of cubes and helped them match them up.

I then showed them an empty bar graph. I asked if anyone had seen them before and if they knew what they could be used for. They gave a variety of answers, but they didn't really know. I told them that a graph is similar to matching things up because the boxes were all equal. I gave each student another pile of cubes and told them to take turns filling in their line. I just had them place the cubes themselves in the boxes by their names (see Figure 8).

I asked them if they could then look at the graph and tell me who had the most. Randall said, "Yes, Magdalena has the most." I asked Salvino how many she had. He counted them and said, "11." Beth said, "You could do it another way. Look at here

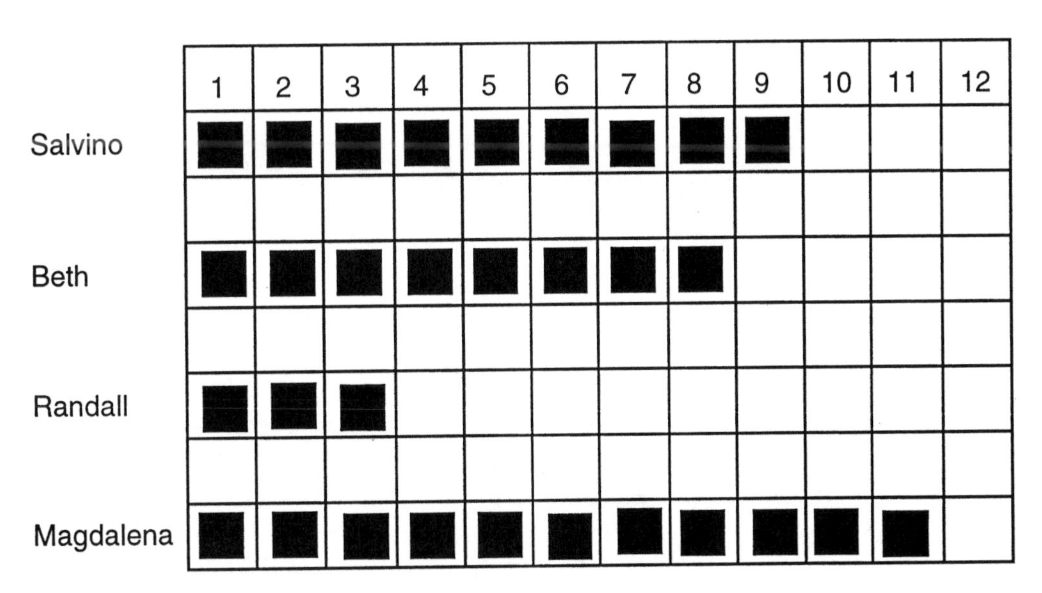

Figure 8 *Introducing a Bar Graph*

(she points to the last cube in the line) and look at this number here (she went directly to the number above)—it's 11." That was exactly what I was leading up to. It was great that Beth showed the rest of the students this instead of me pointing it out. I then asked some questions of the other students trying to get them to use the numbers above the chart as well. Then I asked Beth how many more cubes she had than Randall. She responded, "5." I asked her how she got that number and she said, "I ignored these 3 (indicating her 3 and Randall's 3 that matched up) and looked at mine." I asked her why she ignored those 3 and she replied, "Cause I didn't want to count them." "Why didn't you want to count them?" I asked. She told me "Because they're not part of it." I asked her if they weren't part of what she was counting because they matched up and she said yes. She then told me she counted 2 and 2 which is 4 and 1 more is 5. I asked similar questions of the rest of the students and they showed each other how they got their answers.

I then gave them another chart with four names of teachers and Xs marking how many each had. We went through similar questions with the students rarely making mistakes (when they did say a wrong answer, one of the other kids would usually show them why that couldn't be right or I would ask questions to get them to realize they were making a mistake). With Salvino, I often had to go over how the Xs that matched didn't need to be counted to determine more. If I had 6 cubes and Mrs. Connors had 3—when I asked him who had more he could easily tell me I did and explain why he thought that. But when I asked him how many more I had, he would say 6—counting them all instead of the difference. For the most part, though, all of the students seemed to understand how to read the graph and use it to figure out problems rather well.

As I looked back on the lesson, I think I probably should have done it a lot differently. I think maybe I should have let the students create their own graph at the end. I think I might talk about a vertical bar graph next time and I will probably start off having them draw their own graph to represent something that I give them. I also want to use something besides unifix cubes and I want them to realize that the marks on the graph can stand for anything—cars, sisters, pets, etc. I think bringing in graphs from the paper so that they can see that they are used in the "real world" might be a good idea. Maybe we could use those to figure out information.

The students are working really hard and I know that they enjoy the math time that we spend on Fridays. After only two meetings, three of the children came in straight from lunch and sat down at the table asking, "What do we get to do today?" I also no longer have to solicit other strategies. After someone gives their explanation, I always get, "I have another way!" They are also getting more patient and they really listen to one another. The biggest problem I have is knowing where to go with the lessons. I want to move from the graphing next week to story/word problems. I am wondering if I should give them some word problems connected with the graphs first and then move into just word problems. Is a transition necessary?

I also want to have the students create and solve their own word problems and create problems for a partner as well.

October 7

I tried to observe Magdalena even more closely today than on any other day. She is most challenging because she often says she doesn't know something when I think she does—she is just unsure and she wants to be right.

She can recognize numbers immediately up to 12. After that, she struggles a little and asks questions to help her figure them out ("Is this 21 or 12?"). She can do simple addition (sums under 10) in her head very quickly. I don't ever see her use her fingers and her answer is usually so fast that I wonder if she has just memorized facts. When I ask her to explain how she got that answer, she tells me she just knows, but she does seem to understand the concept as well because she can demonstrate how $2 + 3 = 5$ with manipulatives, and she can use manipulatives to solve problems with larger numbers like $15 + 3 = 18$.

Once Beth explained that the numbers above the graph could be used instead of counting the cubes, Magdalena used that strategy. She easily answered the questions about more and less and could tell what the difference was between two numbers by looking at the graph. Sometimes I think her mathematical reasoning is far more advanced than her verbal skills allow me to see. She cannot always explain how she gets an answer, but she is right pretty consistently. She is still uncomfortable showing the other students how she gets her answers. She likes to solve things quietly—

> not using her hands and not counting out loud. I have to be very careful to try to watch her eyes in order to attempt to determine how she is getting her answers. She is getting a little better about demonstrating some of her methods, however, especially when I put her and Beth in a group together. Magdalena feels more comfortable telling just Beth about something than the entire group.

ASSESSMENT ASSUMPTIONS

This vignette illustrates many of the *Assessment Standards*. First, the teacher believed that assessment should reflect the mathematics that is most important for students to learn. In the first grade, students are expected to take their prior knowledge of counting and rudimentary knowledge of addition, subtraction, and grouping and build on these understandings to develop coherent strategies for solving arithmetic problems. These fundamental concepts are necessary for future work in rational number and algebra. In addition, the ways in which the students solve basic arithmetic problems are inseparable from contextual situations and the models they use to represent them (piles of unifix cubes, bar graphs, etc.). Moreover, because the teacher chose to use a statistical model to illustrate subtraction problems (compare problems; see Table 1), the students have a ready anchor for attaching statistical concepts later on.

Second, it is obvious that the teacher agonized over the different knowledge of her students. She assumed that her assessment should promote equity by giving each student optimal opportunities to demonstrate their mathematical power and by helping each student meet high expectations. This meant, for her, that she had to think simultaneously about the group activity in relation to the individual students' knowledge development. She changed pairs to allow different students to teach each other without one student dominating. She pushed students to develop different strategies to account for different ways of thinking, and she allowed Salvino—who was thinking at a different level than the other children—to contribute significantly to the discussion.

Third, and most important, her approach illustrates the philosophy that assessment should enhance mathematics learning. She did not use her observations (necessarily) to grade students' performances. Rather, she used her observations to design further instruction. She tried different methods, going back to previous experiences and often extending beyond what the students currently knew, in an effort to draw valid inferences about what her students were capable of doing.

Fourth, it is clear from her reflections that she used her assessments as a way of assessing *her own teaching*. Much of what we learn from questioning students and listening to their ways of knowing tells us how well we are doing as teachers.

We can hear your response: "Yes, but she was working with 4 children in basic arithmetic. How would this work with 35 children in a self-contained classroom or with 120 students over six periods?" A fair question for which there is no fair answer. The types of assessments you use as a teacher need to be as varied as the children themselves. The teacher in the vignette made a point of assessing each group of students in her classroom at key junctures in her lessons. Sometimes, she focused on a single group for an entire lesson and rotated groups each day. Sometimes she hit all groups in a single lesson to assess how the whole class was progressing. Sometimes she assigned quizzes as benchmarks to see how students were able to work on their own. Her assessment strategies varied according to the kinds of activities her students engaged in, the grouping strategies she used, and the kinds of information she wanted to obtain.

BENCHMARK OPPORTUNITIES

In addition to the informal, ongoing assessment of students' understanding described in the previous section, there are particular points in time when the activity students are engaged in epitomizes or embodies exactly what we want our kids to know and do. If we could only open a window into their understanding in these activities, we would have a good idea where we have been effective, where we need to spend more time, and where we might go further.

Such benchmark opportunities often occur at the end of a chapter and are often summary activities of the previous series of lessons. Because they summarize a number of important ideas in a single activity, they can be used as a natural assessment task instead of having to write a quiz or exam that will not flow as naturally from the instruction.

For example, the unit *Decision Making* in *Mathematics in Context* (Roodhardt, Middleton, & Burrill, in press) leads 7th-grade students through 4 weeks of exploration in linear systems. The contextual setting for the unit is urban planning—the students must take various constraints and political interests into account to develop an imaginary piece of property. They use coordinate graphing and linear systems to describe, analyze, and make decisions about how compromises can best be made. As a final activity, the students are given a set of new information and are asked to create plans, including scale drawings, of the *best* compromise.

To complete the activity, the students must use the major ideas of the unit in a creative manner. There is no single correct answer—many compromises are *best* given different assumptions. The students must justify their plans using the mathematics in the unit. This activity is a perfect opportunity to assess what the students have learned. It involves all of the major ideas in a meaningful activity, it requires the students to combine the major themes into a coherent justification, and communicate their ideas to others. It also allows for different strategies and approaches.

Not all benchmark opportunities need to be full-blown projects. Sometimes a quick journal entry or question-and-answer session can reassure you that the students are on track and ready to proceed. To be most effective, though, these opportunities should flow from the regular instruction.

ASSESSMENT EVENTS

Tests, quizzes, essays, and other special techniques for determining what students know are also important to include in a balanced approach to assessment. They provide demonstrative evidence that students can do what we say they should be able to, and they are often more formal, organized, and standardized than our observational assessments and benchmark opportunities. We have placed our discussion of these types of assessments at the end of this chapter because they have historically received the major emphasis of most teachers in gaining objective evidence of student learning.

The researcher de Lange (1995) classified the types of tasks that can be included in assessment events into 11 (nonexclusive) categories (see Table 3). In general, the information a teacher can extract from these types of tasks range from low-level reproduction of facts as represented by multiple-choice items to high-level reasoning as represented by fragmented information reasoning tasks, projects, and portfolios.

Please note that all of the types of tasks presented in Table 3 can be valuable for different purposes. Also note that *complexity* does not necessarily refer to *difficulty* of a problem, although the two undoubtedly are related. A multiple-choice item is usually basic, in terms of its complexity, but it can be difficult in terms of the knowledge it accesses. The *Assessment Standards* stress that evidence about students' mathematical performance is needed for a variety of purposes. However, the type and quality of evidence varies with each purpose and with the consequences for students related to each purpose. The point is to design tests that reflect the kind of information you need to make an informed decision about what students' know, can do, and where you are going to lead them.

Sometimes a down-and-dirty test using simple fact-oriented questions is necessary to determine the base knowledge a class has before proceeding to a new unit. At other times, you may want to find out how students are able to connect their understanding of a variety of mathematical

Table 3 Classification of Assessment Tasks (de Lange, 1994)

Type of Task	Example
Multiple-Choice Items	A square can be classified as a: a. rectangle b. parallelogram c. quadrilateral d. all of the above
(Closed) Open Questions	How many quarters does it take to makeup $5.50?
(Open) Open Questions	What combinations of quarters and dimes can total exactly $5.60?
Extended Open Questions	Come up with a rule that describes the relationship between the number of faces, vertices, and edges in solid figures.
Essays	You are members of the city council. The school superintendent pleads, "For every dollar you raise the budget for the police, 20 cents comes from the current education budget! We can't by anymore books!" The police chief claims, "We must have $24,000 more to pay for jail staff!" Using all of the information in your packet, come up with the best possible solution to the problem and defend it using words and pictures.
Oral Tasks	"Susy, let's sit down together. I'm going to let you work a couple of problems, and you tell me your thinking strategies in your own words. If you are unclear about what I am asking, you can ask me for hints."
Two-Stage Tasks	Using your solution to problem 1 (the Open–Open question), create a graph that illustrates the relationship between dimes and quarters for any given total amount.
Student-Generated Tasks	Come up with a couple of percentage problems of your own. Make one an *easy* problem. Write down the problem, show how you would solve it, and describe what makes the problem easy. Make one problem *difficult*. Write down the difficult problem, show how you would solve it (it has to be solvable), and describe what makes the problem difficult (see van den Heuvel-Panhuizen, Middleton, & Streefland, in press).
Fragmented Information Reasoning Tests	FIRTS are *large* problem situations that may have multiple sources of data that interrelate significantly (like real-world problems). They generally involve many days of work and usually need teams of students to approach a reasonable solution. These are more structured than investigations and lead to understanding of students' knowledge of relationships among mathematical domains.

Table 3 Continued

Type of Task	Example
Projects	Using what you have learned about ratio, make a scale model of the automobile of your choice. Describe your measurement techniques and precision, as well as why you chose the scale you did.
Portfolios	Portfolios are not tasks, but they are ways of *organizing* and *presenting* assessment results. They can include information from all of the methods described in this table.

concepts, in which a larger performance assessment of the essay or project type would be appropriate. Neither, however, should predominate in any instructional setting.

WHAT ABOUT GRADING?

All evidence about student performance must be considered as a sample of the possible evidence that could have been gathered. As such, there is considerable potential for error. Multiple sources of information allow for breadth and triangulation to help reduce error. This seems reasonable and logical, but how can teachers do this in the time frame they have available and with a large number of students? In particular, the question arises, "If you accept the philosophy of the *Assessment Standards* and use a variety of assessment alternatives including informal observation records, interviews, quizzes, tests, and performance evaluations, how can you reduce all of this rich information into a letter grade for reporting to parents and the district?"

The grid shown in Figure 9 is a good starting point for expanding your grading system to include the kind of mathematical processes we value. This type of grading scheme is valuable for a number of reasons. It shows the *grade* of any assignment to be a multidimensional profile of strengths and weaknesses across the four process standards in *Curriculum and Evaluation Standards*. Thus, any work you collect from students will reflect their under-

Jacinda Brown

Assignment	Problem-Solving	Reasoning	Communication	Connections
1	4	3	2	3
2	4	4	3	4
3	5	4	4	3

Figure 9 *An Example of a Multidimensional Grading Grid*

standing of the content (represented in Figure 9 by the different assignments) across the four process dimensions in which they are engaged. From this profile, Jacinda seems to be strong in her ability to solve real, ill-structured problems, but her communication and ability to make connections is a little weaker (based on a 1 to 5 scale). The teacher can examine the relative strengths and weaknesses of a child and use this information to bolster weaker areas and communicate strengths to Jacinda and her parents. If a letter grade is required, the teacher can always aggregate the data to produce a single score, but the real emphasis is on helping Jacinda know where to focus her attention in her studies. As in any grading system, care must be taken that the scores children receive reflect what they know and can do.

In a constructivist classroom, assessment ceases to exist as a separate set of tasks imposed upon students. Instead, assessment is a natural reflective activity whereby teachers genuinely attempt to understand the students' unique ways of viewing mathematics, and use these understandings to inform the avenues in which the instruction will progress. By now we hope that you have begun to develop a deeper understanding of the NCTM *Standards*, both in the philosophy in which they were created, and in some of the ways in which they can be implemented practically. Goal Five provides alternative resources that we feel, reflect the *Standards* well, and that will help you teach mathematics better.

SELF-DIRECTED ACTIVITIES: THINKING ABOUT THE *ASSESSMENT STANDARDS*

Standard 1: Assessment Should Reflect the Mathematics That All Students Need to Know and Be Able to Do

If the first four *Curriculum and Evaluation Standards* are the medium by which all students should learn mathematics, then the assessment framework you use to figure out what they know and are able to do should reflect their problem solving, reasoning, communication, and connections.

1 Choose a lesson you currently teach well. Develop an assessment that enables students to demonstrate their knowledge in these ways. Now develop a method of scoring this assessment that would help you determine the standards where students are doing well on that mathematical topic and areas where they are doing poorly. You may assign four different scores based on how well the students do in each of the four standards.

2 Give the assessment task to your students. How well do they do? How can you use your scores to communicate to them and their parents the areas where they need to work harder?

3 Based on the class results, how well did you teach your lesson (Yes, assessment works both ways.) Try to expand this type of assessment to your whole curriculum.

Standard 2: Assessment Should Enhance Mathematics Learning

Consider the following vignette. Imagine you are a 7th-grade student. Forget all of the adult knowledge you have about statistics and think how a 7th grader would approach the following problem:

> The class has just finished a unit on percentages. At the start of the period, the teacher, Ms. Baca, hands out bags full of colored centimeter cubes to each group (groups are about four students each). She says, "Imagine these bags are countries. You are groups of psychologists studying the population of these countries. Your task is to collect samples of data from your country and come up with a reasonable description of the population. But there is a catch. You can't look inside the bag. You can only draw out samples of five "people" at a time and record their characteristics.

After each sample, replace the "people" back into their country. I want you to describe what you think the population looks like after 1 sample of five people, after 10 samples, after 20 samples, and after 30 samples."

How would you start? What knowledge can you, as a 7th grader, bring to this situation? How will you keep track of all this data? How will you communicate your findings to the rest of the class?

The data collected from one group of students is presented in Table 4. Figure 10 presents a record of their thinking.

1 How does this activity help the teacher assess her students' understanding of percentages?

2 What if the students counted instead of using percentages to describe the differences in their samples? How could you, as a teacher, facilitate their discussion to include percentages without telling them to use percentages?

3 How does this activity extend the students' understanding of percentages?

4 How does this activity connect students' understanding of percentages to notions of statistics and probability?

Table 4 Student-Generated Data From the "Countries" Problem

Sample Number	Blue People	Red People	Yellow People	Pink People	Green People
1	2	3			
TOTAL after 1	2	3			
2	2	1	1	1	
3	2	3			
4	1	1	1	1	1
5	2	1	1	1	
6	1	1	3		
7	1	1		1	2
8	2	1	2		
9	3			2	
10	2	1	2		
TOTAL after 10	18	13	10	6	3
11		2		1	2
12	2	2	1		
13	3			2	
14	1	1	3		
15	2	1		1	1
16	1	1	2	1	
17	1		2	2	
18	1	3	1		
19	2	1	1	1	
20	3		1	1	
TOTAL after 20	34	24	21	15	6
21	2	2	1		
22	1	1		2	1
23	2	2	1	1	
24	2	1	2		
25	2	1	1	1	
26	4		1		
27	1	1	1	1	1
28	1	1	1	2	
29	3	1	1		
30	1	2	1		
TOTAL after 30	53	36	31	22	8

After 1 sample:

It looks like there are blue and red people in this country. We don't know how many of each there are, but it looks like there may be more red people since we got 3 but blue was 2.

After 10 samples:

There are more than just blue and red people. We now know there are red, yellow, green and pink people. Now it looks like there are more blue people than the others, but we aren't sure because red was most before, and they are still pretty close. We think the smallest number of people are either pink or green, Cause we didn't get as many.

After 20 samples:

Now we are pretty sure blue is the most and green is the least. There's about 5 times as many blue people as green people. Red seems to be the next most, then yellow, then pink. Green is last. We still don't know how many of each there are because we have a total of 100 people we have drawn, but the bag is too small to hold that many, so we must have drawn some more than once.

After 30 samples:

We really think that our answer after 20 samples is right. Blue is still highest, red is next, yellow is in the middle, pink is next and green is last. But now it looks like Blue is about 6 or 7 times bigger than green. Here is a graph we drew:

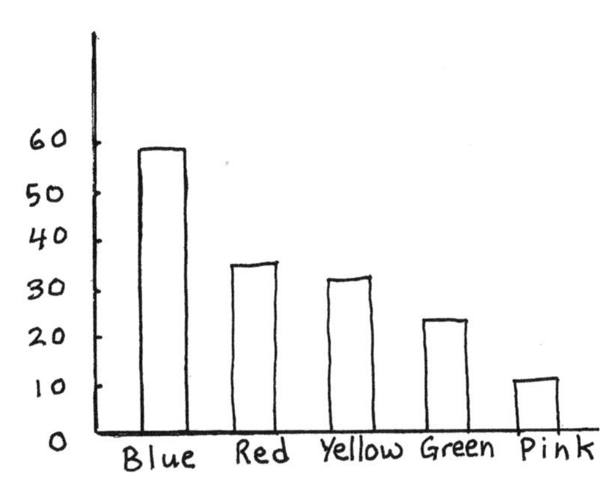

We don't know the exact number of each color in the population, but we think they will be in similar proportions to this graph. Let's say there are 100 people in the population. Then there would be about $\frac{1}{3}$ blues or around 33. That means there would be about 6 or seven times less pinks, which would be about 6. Green is a little more than twice as big as pink, so there would be about 15. Yellow is about 22 and red is about 25, cause it's just a little bit more than yellow.

Figure 10 *Students' Solution to the "Countries" Problem*

Here are some possible teaching extensions. The teacher has several options to pursue after students discuss their reasoning. She can work on ratios and proportional thinking and lead students to calculate the exact proportions their data suggest the people represent; she can draw back and discuss multiple samples and how taking a large number of samples makes estimating the real proportions more accurate; or she can have the class find the average proportions across all countries and see how this compares with the actual number. Each of these options delves into concepts fundamental to understanding statistical and probabilistic reasoning. Calculating the exact proportions is creating a probability estimate ("If I were to draw a single person out of your country, which one would be more likely? If I did this 100 times, how many of each would you predict I would have?"). Discussing multiple samples deals with sampling issues, randomness, error, and convergence ("I saw several groups shaking their 'countries.' Why did you do that?"). Finding the class average deals with using a statistic to best approximate the actual proportions ("What proportions would you most expect to find in the actual population?"). Of course, in any discussion, other issues will arise naturally. ("If we all were sampling from the same 'country,' testing the same 'people,' how come we all got slightly different answers?" "How many samples is adequate?" "What if we took larger samples but the number of samples was less?") This type of simulation draws children into the essential questions statisticians must ask every time they analyze a set of data. It draws children into doing real mathematics in all its complexity.

Extensions of this activity might include doing a real survey to determine characteristics (interests, attitudes, demographics) of children in the school. Students can develop a survey, create an appropriate sampling frame, and find out interesting things relevant to their immediate lives. Writing, illustrations, and other communication methods can be used to report findings to the class, the student body, or to the school board.

A Gratuitous Note on Why We Like to Teach Statistics

One of the values of introducing statistical ideas into the elementary and middle-school classroom is their connections to other mathematical content. Children gain valuable practice in arithmetic as they calculate ratios, proportions, add, subtract, multiply, and divide numbers to obtain useful statistics. Developing statistical formulas ties students' work to algebraic thinking. Using graphs and other proportional models links their thinking to geometric notions. Statistics and probability are mathematical areas that make full use of knowledge in the other areas. Moreover, statistics, apart from being a field of mathematics, is essential to understanding other content areas such as science, social studies, and journalism. The opportunity for curriculum integration is powerful as you explore how other fields use statistics.

Standard 3: Assessment Should Promote Equity

The *Curriculum and Evaluation Standards* stress that the mathematics we envision is for *all* students. This means that we as teachers need to assume that all of our students can learn mathematics—and learn mathematics well, regardless of race, gender, physical abilities, background, socioeconomic status, or other reasons we traditionally have excluded people from playing the game. This does not mean that all students learn mathematics in the same way, nor does it mean that all students learn mathematics to the same level for any given activity. What it does mean is that everybody has intelligent ways of making meaning out of mathematical situations, and that the differences in ways of thinking allow all of us to think about mathematics in

Los siguientes diagramas representan las fracciones 5/5, 4/5, 3/5, 2/5, 1/5.

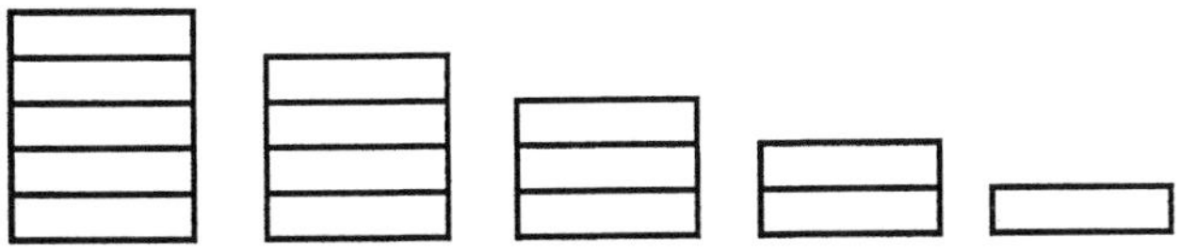

Illustra el siguiente problema de division con el dibujo.

¿ Cuantas veces cabe 2/5 en 4/5?

Figure 11 *An Assessment Problem on the Division of fractions (with appreciation to Dr. Alfinio Flores, Arizona State University)*

a more flexible, powerful way. It means that we have to take the cultural, cognitive, physical, and emotional backgrounds of our students into account when designing assessments, so that all students have the opportunity to show us what they know and can do.

Try assessing your knowledge of fractions in Figure 11. For many of you this might prove to be a difficult problem despite understanding division of fractions. First, you may have been baffled when encountering a different language than English. Second, you may not have caught on that the problem asks you to illustrate your answer with a drawing. But you may have noticed the diagram above the problem gives representations of the fractions involved. How could the diagram help you understand the problem?

1 Come up with some ideas that you could use to help an English-speaking student show what they know about fractions using the problem in Figure 11.

2 Extend your thinking to develop strategies for allowing students whose primary language is not English to display their understanding of fractions.

Gardner (1983) outlined a theory that maintains that people have at least seven different ways of thinking about the world (*multiple intelligences*). Few people are outstanding in all ways of processing information: We all have our preferred modes. Some people are adept at visual and artistic thinking, whereas others are more adept at logical/mathematical thinking. (Note: We do not believe that Gardner's domain of logical/mathematical thinking is synonymous with *thinking mathematically* as envisioned by the *Standards*. Thinking mathematically may, and probably does, use more than logical/mathematical thinking as defined by Gardner, and many nonmathematical problems use logical thinking.)

3 Take the following list of modes of thinking and come up with ways all students can make sense of a mathematical topic *at the same time!*

- Visualizing and drawing (spatial intelligence),
- Manipulating and building (spatial and kinesthetic intelligence),
- Using writing and verbal argument (linguistic intelligence),
- Patterning in numeric as well as geometric ways (logical intelligence),
- Strategizing and group work (logical and interpersonal intelligence).

Standard 4: Assessment Should Be an Open Process

This assessment standard asserts that all people affected by assessment should be privy to the purposes, processes, and stakes involved in obtaining information about student learning. This first and foremost assumes that students should be "informed about what they need to know, how they will be expected to demonstrate their knowledge, and what the consequences of assessment will be" (NCTM, 1995, p. 17). In addition, students should receive feedback about their work in a timely, thoughtful fashion. This, in turn, contributes to their future learning.

Before engaging in any assessment task, developers of *The Packets Project* (Educational Testing Service, 1995) asserted that for students to be successful at mathematics they must understand the following questions:

1

Who wants to know? This alerts students to the audience their work is targeted toward. In a statistical problem, for example, knowing whether the audience is a group of environmentalists or a group of developers—and understanding that these groups are not mutually exclusive—determines the nature of what could be considered a reasonable set of conclusions.

2

What am I going to produce? Students should understand what is expected. The products of assessment can include a report, a model, a picture, a verbal conjecture or argument, a symbolic expression, or other mode of communicating the students' reasoning. In the hypothetical statistical example mentioned previously, the product might be a report to the Environmental Protection Agency on the impact of leach-field sewage systems on stream ecology, or it might include a graph of changes in plant and animal life as a result of increased sewage output.

3

What is the purpose of the product? Should students target a lay audience? If so, the purpose of the product may be to inform. Should students target a governing agency? If so, the purpose of their work may be to make recommendations. Should students target the scientific and mathematical community? If so, the purpose may be to create a new method and argue its validity. By understanding the reasons for creating a worthwhile product, students will be more motivated to do good work and will be better able to focus on the important issues.

- Take a look at the assessment tasks you give your students. Can the students reasonably answer these questions? If not, how can you build this kind of openness into your assessment practices?

- Create a new assessment task that enables students to problem solve with the three questions to help them focus their work. What does this activity reveal to *you?*

Standard 5: Assessment Should Promote Valid Inferences About Mathematics Learning

To make valid inferences about what a child knows mathematically, the assessment task must mirror, at least in part, the ways in which the child learned the mathematics initially. This implies that if a child learns in a social environment, they must also at times be assessed socially.

1 How can you make accurate assessments of what children know and can do when they are assessed as members of a team? Box 3 provides an assessment tool (a test) the first author often uses to find out what his students know about rational numbers in many of their nuances. Critique this assessment task. (Be brutal!) How does the partner section lead the student to the individual section? How can students use multiple ways of thinking about ratios to solve each of the problems?

Box 3 A Test of Rational Number Understanding

Instructions: Do your best to solve each of the following problems. You may use any of the notes you have taken in class to help you. The first part of the test is to be done with a partner. The second part is to be done on your own.

Part 1: Partner Math

1. Save-Mart is having a 1/2-off sale. That means every item in the store will be sold at half its original price.
   a. What percentage discount does 1/2 off represent?
   b. What will be the sale price of a $45 sweatshirt?
   c. Come up with a general way to express the relationship between the original price and the sale price for *any* item.
2. Tracy and Rose are racing their bicycles down Silly Mountain Road. The downgrade is an 8% slope. This means that for every 100 meters Tracy and Rose travel in the horizontal direction, they drop 8 meters in the vertical direction.
   a. Draw a picture that illustrates this slope.
   b. If Tracy goes 225 meters, how far has she dropped in elevation?

Continued

c. Let's say it is 5 kilometers to the ocean by way of Silly Mountain Road. If the grade stays roughly the same, at what elevation are Tracy and Rose now?

d. In general, describe what happens (both mathematically and to Tracy and Rose) when we change the slope of the road.

3. A forensic scientist has measured the humerus bone and height of 10 cadavers to determine if there is a relationship between the length of the humerus bone and the height of a person. The results are tabled:

Humerus (cm)	Height (cm)
26	152
34	173
37	192
25	159
35	180
22	132
23	142
16	120
14	108
10	101

a. About what height is a person with a humerus bone that is 25 cm?

b. About how tall is a person with a humerus bone that is 12 cm?

c. Plot the line that best fits this data. How well does the line predict the heights of the cadavers?

d. Suppose the scientist found a giant humerus of 72 cm. Describe how you would determine the height of the giant. Extend this description to find the height of any person.

Part 2: Individual Math

4. Remember all of the things we have done over the past 4 weeks in ratio, geometry, algebra, and statistics. Using the problems above as examples, describe at least one of the underlying mathematical themes. (Hint: What do all of these problems have in common?)

You may have noticed that the questions asked in problems 1 through 3 are structured in a specific way. Research on cognitive development and assessment has revealed that if assessment tasks are structured in such a way that prior questions within a problem can be solved using more rudimentary thinking, and consequent questions must be solved with progressively more difficult thinking, the assessor (teacher) can better determine where students are in their development of mathematical thinking about a particular concept (Collis, Romberg, & Jurdak, 1986). The Structure of Learned Outcomes (SOLO) taxonomy is a valid, coherent way of determining the level of thinking students are able to produce in a mathematics task. Collis et al. have suggested that subitems in a problem can be structured in five levels:

Level 0 *Prestructural.* A student is unable to make sense of the information in a problem in any useful way.

Level 1 *Unistructural.* A student can use one obvious piece of information coming directly from the problem stem.

Level 2 *Multistructural.* A student can use two or more separate pieces of information coming directly from the problem stem.

Level 3 *Relational.* A student can use two or more pieces of information from the problem stem in an integrated fashion (put together or use conjointly).

Level 4 *Extended Abstract.* A student can use an abstract general principle derived from or suggested by information in the problem stem.

With this type of structure, a student who is at an extended abstract level of thinking will be able to answer all of the questions in a problem. A student who is at the multistructural level, however, will be able to successfully answer only those questions that are written at a multistructural or unistructural level.

1

Problem 4 general adheres to the SOLO structure. Examine each of the subitems and describe how they fit the SOLO taxonomy. Examine one of your assessment tools; modify it to fit this developmental structure better. Experiment with the modified tool and see how it helps you determine your students' levels of thinking.

Standard 6: Assessment Should Be a Coherent Process

The coherence standard implies that just like all aspects of a mathematical curriculum should fit together to create a meaningful set of experiences that lead to mathematical understanding, so must the assessment program fit together to provide information in a meaningful way for teachers, administrators, parents, and others, so that student learning is enhanced. The writers of the *Assessment Standards* stress *balance* in the types of assessment activities we design for our students. Paper-and-pencil tasks are fine for some purposes, but oral tasks are better for others. Group projects can provide insight into the cooperative work of a class but are much harder to assess for individual knowledge. The teacher needs to build in a multitude of strategies and approaches for determining what their students know. (This applies to district-wide and state-mandated tests as well.)

1

In a team, examine the assessment program for your school or district. Does it provide coherent information about all important aspects of the curriculum? Where are the holes? Does it value multiple strategies? Does it allow *all* students to show what they know?

2

Examine your own set of assessment strategies. Does your strategy provide you with the information you need to teach in the way envisioned by the *Standards*? Does any single type of assessment predominate, or do you use a balance of methods?

goal five

Finding Resources to Help

This goal presents a number of resources that you can tap into to help you change the nature of your mathematics classroom. In addition, we present a few useful points to keep in mind when dealing with parents, administration, and other teachers. This is a starting point that we hope stimulates you to implement the *Standards* more fully.

To assist schools in certifying teachers who have the knowledge and skills needed to implement effectively the reform vision outlined by the NCTM *Standards*, the Council of Chief State School Officers (1995) have presented the paraphrased principles for *all* teachers of mathematics in grades K through 12 in Box 4.

Box 4 **Model Principles for Beginning Teachers of Mathematics (CCSO, 1995).**

1. Teachers need to understand the important ideas of mathematics—and the processes and perspectives on which these ideas are built—in terms of the whole K–12 curriculum.
    a. Teachers must understand mathematical ideas from the following areas:
        - number systems and number theory,
        - geometry and measurement,
        - statistics and probability,
        - functions, algebra, and concepts of calculus,
        - discrete mathematics.

    b. Teachers need to develop their knowledge of mathematics through the following processes:
        - problem solving in mathematics,
        - communication in mathematics,
        - reasoning in mathematics,
        - making mathematical connections.

    c. Teachers need to develop the following perspectives about mathematics as a discipline:
        - the history of mathematics,
        - mathematical world views,

- mathematical structures,
- the role of technology and concrete models in mathematics.

2. Teachers of mathematics need to understand how children learn and develop so that they can provide learning opportunities that support the personal, intellectual, and social development of their students.
3. Teachers of mathematics need to understand how students differ in their approaches to learning and to create opportunities for all students to learn mathematics.
4. Teachers of mathematics need to use a variety of strategies to encourage students' development of mathematical thinking, problem solving, and performance.
5. Teachers of mathematics need to understand individual and group motivation to create an encouraging learning environment.
6. Teachers of mathematics must use effective verbal, nonverbal, and media communication techniques to foster inquiry and collaborative interaction in the classroom.
7. Teachers of mathematics need to plan their instruction based on their knowledge of subject matter, their students, their community, and their curricular goals.
8. Teachers of mathematics must use both informal and formal assessment strategies to ensure the continuous intellectual, social, and physical development of their students.
9. Teachers of mathematics must be reflective practitioners who continually evaluate their teaching of their students and the community and who actively seek out opportunities to grow professionally.
10. Teachers of mathematics must foster collaborative relationships with others in the community to support students' learning and well-being.

This list of principles may seem daunting to you. It does to us. But, the argument being made is a logical one. Teachers of all subjects should understand the history, systems, concepts, and processes of their fields to ensure that the content being taught is both coherent and important. Teachers need to understand their students—how they learn and how they develop—so that instruction is tailored to the particular needs of these students. Teachers should understand important pedagogical principles so that content, instruction, and assessment motivate and involve students. Teachers need to be connected to the community so that parents, administration, and other teachers can work together toward the common goal: educating our children. As we stated before, do not try to change everything all at once. There are certain areas in which you feel you are doing well and others in which you feel a bit fuzzy. Choose the principle in which you feel you need the most work. How can you get the information you need to accomplish this? How can you develop these skills into your teaching repertoire? Who can you call to help you in your quest?

This chapter provides a list of practical resources that will assist you in making the transition toward the mathematics teaching the *Standards* envision. This list is by no means exhaustive. There are a multitude of good resources you can tap into in your quest for change, so keep an eye out.

INNOVATIVE CURRICULUM PROJECTS

It is silly, we think, to assume that any individual teacher will be able to implement all of the *Standards* in a classroom setting with no support. In the curricular arena alone, developing a coherent system of instruction for an entire year can be overwhelming without tapping into resources that have been designed explicitly to conform to a new vision of mathematics and its learning. The following projects are perhaps the most innovative curricula available at this time. Most are still under development, but many have released units of work or trial units in prepublication form for teachers to use in their classes. We will provide a brief description of each project, along with ways in which you can contact a project representative or publisher for more information. We encourage you to contact the people involved in each project in your grade level. We have been an integral part of one of the middle grades projects (*Mathematics in Context*), and we have interacted with key people in each of the other projects. We truly feel that the curricula these projects produce will give teachers *real* choices.

Elementary Projects

The Cooperative Mathematics Project

The Cooperative Mathematics Project (CMP) develops instructional materials for grades K through 6 that support students' mathematical and social development. This project's program is entitled *Number Power*. It takes a constructivist approach to teach by providing investigations in which students in pairs or groups of four discuss and write about their thinking and solutions. Students develop and apply their understanding of numbers through estimation, data analysis, and by creating their own computation and problem-solving strategies.

The CMP also focuses on students' social development. Social benefits of the program include students learning what it means to be responsible, fair, and concerned for others. Students learn how to analyze the effects of their actions on themselves and others and on their work. Moreover, students learn to work together, reflecting on successes and relating them to effective group skills.

There is a *Number Power* teacher resource book for each grade level. Each book contains three multiweek units. For more information on CMP, contact the publisher:

Addison-Wesley Publishing Company
200 Middlefield Road
Menlo Park, CA 94025
(415) 854-0300
Internet: http://www.aw.com/

Everyday Mathematics

Everyday Mathematics is the elementary component of the University of Chicago School Mathematics Project (UCSMP), whose secondary materials are published currently by Scott Foresman. This project has expanded on the successful Everyday Mathematics program K–3, which is already available. The project is creating a curriculum for students of normal ability range to succeed in more ambitious mathematical concepts and skills, regardless of gender, race, socioeconomic status, and other factors that have traditionally been discriminated against in mathematics education. Much of the success of this program is attributed to the high interest level of the mathematical and contextual features of the activities.

The design of instruction involves introducing mathematics at a playful level up to 2 years before more formal instruction. This provides a base of experience with mathematical ideas so that students feel comfortable taking them to the next level. Students make extensive use of measurement and mathematical modeling incorporating a variety of tools including manipulatives, calculators, and computers. Mathematical structures and algorithmic procedures are developed as one important aspect of mathematical modeling. The emphasis on modeling requires students to become involved in data collection, analysis, and representation, with technologies used as tools to help them solve problems. Most work is done with partners or small groups, with instructional strategies balanced around discussion, explorations, projects, and exposition by the teacher. The units come with a high level of teacher support, tools, and ongoing assessment. To contact the Everyday Mathematics project, contact the publisher:

Everyday Learning Corporation
1007 Church Street
Evanston, IL 60201
(800) 382-7670

Investigations in Number, Data, and Space

Investigations in Number, Data, and Space is designed for grades K through 5. It is organized around a set of units that take approximately 2 to 4 weeks, involving students in investigations that explore major mathematical ideas. Through these investigations, children develop fluency in using mathematical tools and concepts to solve problems, flexibility in the ways in which they approach problems, and proficiency in evaluating the results of their investigations. A great deal of emphasis is placed on communication using a variety of verbal, symbolic, and manipulative modes. The problem contexts require students to share their experiences from their families, cultures, and communities in order to break down barriers that exclude some students from successful participation in mathematics.

The Investigations project has four major goals:

1. Offer students meaningful mathematical problems,

2. Emphasize depth as opposed to surface-level exposure to mathematical ideas,

3. Teach students to communicate mathematically with peers and the teacher,

4. Ensure a much larger proportion of students become mathematically literate.

As students become involved in meaningful mathematical experiences, they spend more time exploring problems in depth, finding more than one solution to many of the problems they attempt. They invent their own strategies instead of relying on memorized procedures, and they choose from a variety of concrete manipulatives and technologies to help them in their work. Communication is developed through drawing, writing, and discourse. Children work in pairs, small groups, individually, and as a whole class. Balance of grouping strategies is stressed. For more information about the Investigations in Number, Data, and Space project, contact the publisher:

Dale Seymour Publications
P.O. Box 10888
Palo Alto, CA 94303
(800) 417-2321
Internet: http://aw.com/dsp/

Transitions in Mathematics and Science

The Transitions in Mathematics and Science (TIMS) program is designed for all students in Grades 1 through 5. It is based on the belief that scientific investigation of real-world situations is the ideal context for learning mathematics. The program, entitled *Voyager: A Mathematical Journey Using Science and Language Arts*, is an integrated system that focuses on mathematics but creates explicit links to science and language arts curricula. All aspects of mathematical content are introduced at each grade level, including measurement, statistics, mental arithmetic, graphing, geometry, ratio, simple algebra, estimation, and patterns and relationships. To facilitate students' learning of such challenging content, lessons are developed in everyday experiences, in an attempt to ensure success for all.

Within an investigation, children are led through four phases of scientific method: (a) drawing a picture, (b) collecting and organizing data, (c) graphing the data, and (d) analyzing the experimental results. This forges the necessary connections between worldview, method, variables, and mathematical models to create a natural flow of scientific thinking. As students move through the curriculum, ideas move from simple to complex, building the foundation for proportional reasoning, multistep logic, and algebraic thinking. Reading, writing, and language arts are used to communicate collaboratively in scientific teams.

The Voyager program includes a balanced package of assessment ideas, tools, and tasks. These assessment strategies are designed to discover the level of students' understanding, and they are also meaningful learning experiences. Whole-class instruction, small-group projects, and individual work all have their place in the program. For more information on Voyager and the TIMS program, contact the publisher:

> Kendall/Hunt Publishing Company
> 4050 Westmark Drive
> Dubuque, IA 52004
> (800) 228-0810

Middle-School Projects

The Connected Mathematics Project

The Connected Mathematics Project is funded by the National Science Foundation to develop a complete mathematics curriculum for Grades 6 through 8. This project builds on the success of the earlier project, the Middle

Grades Mathematics Project, with input from other successful programs. The philosophy underlying the curriculum is that *how* students learn shapes *what* they learn in mathematics. This translates into five instructional themes: (a) understanding, (b) connections, (c) mathematical investigations, (d) representations, and (e) technology. These themes guide the kind of discourse necessary to support student growth in mathematical understanding.

The project emphasizes the connections between mathematics and other disciplines such as science, social science, business, as well as within mathematical domains. In addition, connections are made between mathematics and the unique interests and abilities of middle-school students and between elementary and secondary school experiences. Special attention is paid to the meaningful interdependence of teaching, learning, and assessment. Each unit uses problem solving as defined by the *Curriculum Standards* to develop a major concept in the following content areas: number, geometry, probability, statistics, measurement, and algebra. Each grade level has eight units that are designed to take approximately 4 to 6 weeks to accomplish. The *Connected Mathematics* Project materials are published by Dale Seymour Publications:

Dale Seymour Publications
P.O. Box 10888
Palo Alto, CA 94303
(800) 417-2321
Internet: http://aw.com/dsp/

Mathematics in Context: A Connected Curriculum for Grades 5 through 8

The Mathematics in Context (MIC) project is designed to create a comprehensive mathematics curriculum for the middle grades that reflects the content and pedagogy suggested by the NCTM *Curriculum and Evaluation Standards for School Mathematics*, and *Professional Standards for Teaching Mathematics*. The project uses the developing knowledge in the field of mathematics education and the prior experience of curriculum developers in the Netherlands to create a research-based, teacher-oriented set of materials that will form a prototype of school mathematics for the 21st century. The goal of the MIC curriculum is to provide materials that enable all students to do mathematics well.

The development of the curriculum units reflects a collaboration between research and development teams at the Freudenthal Institute at the University of Utrecht, The Netherlands; research teams at the University of Wisconsin; and a group of middle-school teachers. A total of 40 units have been developed for Grades 5 through 8. These units are unique because they make extensive use of realistic contexts in science, social studies, and everyday life. Units emphasize the interrelationships between mathematical domains such as number, algebra, geometry, and statistics and their connections to the real world.

Because the philosophy underscoring the units is that of teaching mathematics for understanding, the curriculum will have tangible benefits for both students and teachers. For students, mathematics should cease to be seen as a set of disjointed facts and rules. Rather, students should come to view mathematics as an interesting, powerful endeavor that enables them to understand their world better. *All* students should be able to reason mathematically. Thus, activities will have multiple levels so that the more able students can go into depth and the students having trouble still can make sense out of the activity. For teachers, the rewards of seeing students excited by mathematical inquiry, their redefined role as guide and facilitator of inquiry, and their collaboration with other teachers should result in innovative approaches to instruction, increased enthusiasm for teaching, and a more positive image with their students and society at large.

The project has developed an entire mathematics program—40 units, teachers' guides, assessment materials, a technological platform, and staff development materials—with the resources and support necessary for successful implementation. For each grade level there will be a grade overview document, a complete set of 10 units; a set of blackline masters, *Number Tools* that revisit important number concepts and can be used to supplement other units as necessary; and *News in Numbers*, a series of activities based on newspaper articles that can be used to develop estimation skills and number sense as well. For information, contact the publisher:

Encyclopaedia Britannica Educational
Corporation
310 South Michigan Avenue
Chicago, IL 60604
(800) 554-9862
Internet: http://www.ebec:com/

Middle-School Mathematics Through Applications Project II: The Comprehensive Program

The Middle-School Mathematics Through Applications Project II: The Comprehensive Program (MMAP II), developed by researchers at the Institute for Research on Learning, builds on the success achieved in the original Middle-School Mathematics Through Applications Project (MMAP). MMAP II will result in a comprehensive, 3-year mathematics program for sixth-, seventh-, and eighth-graders. At the same time, the Institute for Research on Learning continues to build a broad alliance of math educators and math teachers interested in applications-based mathematics. This alliance will become an active community that cooperates in developing MMAP materials and supporting teachers' efforts to use them in classrooms.

The Goals of MMAP II include

- Developing materials necessary to create a complete, flexible, and innovative math program that is capable of serving all sixth-, seventh-, and eighth-graders, regardless of their prior skill and ability levels,
- Providing math educators necessary resources, materials, and guidance to assemble MMAP II materials into a coherent, customized, and affordable mathematics curriculum that meets specific local needs,
- Building guidelines and activities that help teachers (a) identify mathematics embedded in students' work and (b) systematically evaluate student mathematical competencies,

- Galvanizing a unique and growing community of math educators, researchers, math-using professionals, and others who collaborate in identifying the needs and means necessary for developing MMAP II materials.

MMAP II will create innovative software and investigations to simulate real-world problems that rely on math concepts and skills to solve. Piloting of similar materials developed during the original NSF grant revealed this approach to be extremely motivating for students of diverse ethnicities and abilities, prompting them to engage in substantive mathematical issues and problems at high levels. The project uses computer technology in integrated, nontrivial ways to model students' solutions to complex problems.

MMAP II uses an interactive, cyclical research and development process. In addition, the project continues to establish numerous partnerships and collaborations with math-using professionals, schools, school districts, teacher development initiatives, universities, and preservice teacher education programs.

For more information, contact

Jennifer Knudsen
MMAP Associate Director for Curriculum and Instruction
Institute for Research on Learning
66 Willow Place
Menlo Park, CA 94025-3601
(415) 614-7900

Seeing and Thinking Mathematically

The Seeing and Thinking Mathematically (STM) project is designed to provide students with experiences that center around the importance of mathematics in the human experience. Students use mathematics to design and build, create and explore, analyze and formalize, to create an understanding of the central ideas of mathematics: patterns, representations, proportional reasoning, functions, and mathematical models. Students actively explore mathematical ideas using physical and pictorial models as they explore investigations and projects.

This project is concerned with developing habits of mind in students: visualizing, representing, calculating, computing, modeling, inventing, proving, systematizing, and communicating about mathematics. STM provides methods for students from different backgrounds and experiences to use and build on their own strengths to become more mathematically powerful. Each unit provides a rich set of resources for teachers to adapt to their classrooms and teaching styles. Each unit also provides vignettes of experiences real teachers have had teaching the unit. Assessment alternatives including portfolios, scoring rubrics, and journaling are incorporated naturally in STM activities. All STM activities are embedded in relevant contexts so the mathematics always has a purpose. For more information, contact the publisher:

Heinemann
361 Hanover Street
Portsmouth, NH 03801-3912
(800) 541-2086

Six Through Eight Mathematics

The Six Through Eight Mathematics (STEM) project is designed to provide teachers with mathematically accurate, technologically appropriate curricula that connects mathematical concepts with the sciences and other mathematical fields. STEM materials are problem-centered, thematic modules developed around mathematical applications. Projects and investigations may last several days, and many involve work outside of class time. Students are expected to work cooperatively in solving problems.

A number of unifying concepts will tie the units together mathematically and conceptually. Among these concepts are proportional reasoning, multiple representations, patterns and generalizations, and mathematical modeling. The goal of the project is to show students that mathematics is both exciting and useful. Students in the program will develop communication skills and flexible approaches to solving mathematical applications. Computer spreadsheets and scientific calculators are integrated throughout the curriculum. For more information, contact the publisher:

McDougal Littell/Houghton Mifflin
222 Berkeley Street
Boston, MA 02116-3764
(617) 351-5000
Internet: http://www.hmco.com

High School Projects

The Interactive Mathematics Program

The Interactive Mathematics Program (IMP) has created a 4-year, problem-based high school mathematics curriculum designed to meet the needs of both college-bound and non–college-bound students.

Basic assumptions and features of the project include (a) a shift from a skill-centered to a problem-centered curriculum; (b) a broadening of the scope of the secondary curriculum to include such areas as statistics, probability, and discrete mathematics; (c) changes in pedagogical strategies, including emphasis on communication and writing skills as well as access to appropriate technology; and (d) expansion of the pool of students who receive a core mathematics education.

IMP replaces the sequence typical of most high school mathematics programs with an integrated course. Whereas much traditional mathematics teaching is structured around skills and concepts in isolation, IMP calls on students to work in context, experiment with examples, look for and articulate patterns, and make, test, and prove conjectures.

IMP differs from other problem-based instructional programs in that the curriculum is structured around a series of units, each organized around a central problem or theme. Students develop skills within the context of these large, complex problems. Over the course of 4 years, the curriculum revisits themes and concepts in a spiraling sequence, providing students with ongoing opportunities to develop mathematical understandings with increasing sophistication.

Opportunities for assessing learning include self-assessment, student portfolios, oral presentations, written explanations, teacher observations, and group work, as well as end-of-unit and end-of-semester tests.

To give students experience in working independently on substantive problems, IMP has incorporated Problems of the Week (POWs) into each unit. Some POWs are related directly to the specific mathematics of the unit under study; others are classic mathematics problems, independent of the unit.

The interactive aspect of IMP refers, in part, to the program's emphasis on students interacting with each other by working in groups. Together, students tackle problems that usually are too complex to be solved by any one individual. Students make written and oral presentations that help clarify their thinking and refine their ability to communicate mathematically. Although students are allowed to collaborate on problems, evidence of individual interpretation can be seen in their written portfolio work. Further information can be obtained from the IMP National Office:

Interactive Mathematics Program
6400 Hollis Street
Emeryville, CA 94608
(510) 658-6400
e-mail: imp@math.sfsu.edu

or from the publisher:

Key Curriculum Press
P.O. Box 2304
Berkeley, CA 94702
(510) 548-2304
Internet: http://www.keypress.com/

Applications Reform in Secondary Education

Applications Reform in Secondary Education (ARISE) is a secondary curriculum development project funded by the National Science Foundation to the Consortium for Mathematics and Its Applications (COMAP). The project is designed to develop a 9th- through 11th-grade mathematics core curriculum that focuses on the applications of mathematical ideas.

In the ARISE project, the mathematics students learn *arises* from the applications in which mathematics really is used in the real world. The units are not built around mathematical topics per se but center instead on application themes such as "Decision Making in a Democracy," with the mathematical topics occurring as strands within each unit. Students analyze real-world situations, develop mathematical models to fit them, check their models against reality, and modify them. This approach results in a natural integration of mathematical topics that avoids the isolated treatment mathematics has endured in the past. ARISE uses technology, including graphing calculators and computers, whenever they are appropriate in solving a particular problem. Video technology also is used to situate the real-world contexts for students.

The project is developing materials for teachers and students. There are explorations suitable for group learning as well as individual homework and assessments. For more information, contact COMAP:

> Consortium for Mathematics and Its Applications
> 57 Bedford Street
> Suite 210
> Lexington, MA 02173
> (617) 862-7878

The Core-Plus Mathematics Project

The Core-Plus Mathematics Project (CPMP) is a project designed to develop a core mathematics curriculum for the first 3 years of high school (Grades 9–11) for all students, with an option for the 4th year (Grade 12) intended to make the transition to college mathematics. The underlying philosophy of CPMP is "mathematics as sense making." It is grounded in the following principles:

- A comprehensive mathematics curriculum should feature multiple strands of algebra and functions, geometry and trigonometry, statistics and probability, and discrete mathematics, all connected by common contexts and through the themes of *data, representation, shape, change, and chance.*

- The mathematics must be accessible to all students, with differences in background, interests, and performance accounted for by the depth and level of abstraction to which topics are pursued, and by the types of activities used and the difficulty of their applications.

- Activity must focus on meaning, not mechanics. Graphing calculators should be integrated into the curriculum to allow students to be able to focus on mathematical thinking and not on mere arithmetic or symbol manipulation.

- The curriculum must emphasize mathematical modeling through data collection, representation, prediction, and simulation.

- Instructional practices should emphasize cooperative as well as individual problem solving.

- Assessment of mathematical thinking and performance should be multidimensional.

The strands of algebra and functions, geometry and trigonometry, statistics and probability, and discrete mathematics are developed each year in the program. Important mathematical ideas are revisited continually at higher levels so that students can develop a fuller understanding of mathematics. The project makes use of the graphing calculator, and a resource package assists the teacher in effectively translating the units into action. The project is being published by

Jansen Publishing.
450 Washington St.
Suite 107
Dedham, MA 02026-4449
(800) 322-6284

For more information, contact:

Core-Plus Mathematics Project
Department of Mathematics and Statistics
Western Michigan University
Kalamazoo, MI 49008
(616) 387-4562

Math Connections: A Secondary Mathematics Core Curriculum

The goals of the Math Connections project are to increase mathematical power to all students and stimulate the participation in college preparatory mathematics of minorities. This project is tied to a restructuring and professional development project in the Hartford, Connecticut, school district. The curriculum will incorporate technology and alternative assessment techniques to provide computational power and valuable insight into students' thinking for teachers.

The curriculum will be developed in modules designed around a semester, rather than a year-long course. Emphasis will be placed on investigation, team work, and written and oral communication. Connections will be made within units, across units, and across courses in the high school mathematics experience. Students will actively study patterns and develop multiple representations of mathematical phenomena using graphing calculators and computers. They will work with real-world data to emulate life and job-related experiences. The classroom model will involve students in becoming more active constructors of their own knowledge, with a greater emphasis on self-evaluation. For more information about the Math Connections project, contact the Connecticut Business and Industry Association:

Math Connections
CBIA Education Foundation
370 Asylum Street
Hartford, CT 06103-2022

Systemic Initiative for Montana Mathematics and Sciences

The Systemic Initiative for Montana Mathematics on a Science (SIMMS) is a dual project designed to implement systemic change in the mathematics experiences of Montana students and to create a comprehensive secondary curriculum appropriate for students across the nation. The goals of the project are

- Redesign the 9–12 mathematics curriculum for all students using an interdisciplinary approach,
- Develop curricular and assessment materials consistent with the first goal and with the NCTM *Standards*,
- Incorporate the use of technology into all levels of mathematics education,
- Increase the participation of females and Native Americans in mathematics and science, and increase the number of females and Native American teachers in mathematics and science,
- Redesign the certification and recertification standards for teachers,
- Develop an extensive in-service program in mathematics for grades 9–16 to prepare teachers to implement the integrated program,
- Develop a support structure with the legislature, public, and teachers to implement the program effectively.

Although many of these goals are specific to the Montana State Systemic Initiative, all of these goals work together to inform the curriculum development component of the project.

The SIMMS curriculum covers topics from a wide range of mathematical fields and connects those topics to emphasize the unity of mathematical ideas. The materials are organized into 6 levels of approximately 16 modules each. The first two levels are year-long courses expected of all students, and the other four levels are options students can take according to their interests and goals.

For more information on SIMMS, contact the project office:

SIMMS Project
University of Montana
Math Buidling Room 7E
Missoula, MT 59812

ASSESSMENT

The Packets Program

The Packets Program is a learning- and performance-based assessment package for the middle grades. This project is unique because the assessment situation is also a rich learning environment. The activities are built around newspaper articles that present real-life dilemmas, facts, and mathematical data. The program uses a three-part model. During the warm up, students read the newspaper articles and answer questions that lead them to focus on the issues at hand in an effort to get the students to go beyond the articles and project the big ideas. Then students work in groups on a focus project. Each group creates a written solution to questions that focus on the big ideas developed in the warm up and orally present their solutions to the class. The features of the focus project are an open-ended format; cooperative learning; connections to science, social studies, and other disciplines; and extensive use of problem-solving and communication skills. Closure activities require children to revisit the math concepts developed in the warm up and focus project and go into more depth by applying what they know in a new context. During the entire process, students *learn* while they are being *assessed*.

The Packets Program supports the assessment process by providing

- Criteria for evaluating student work,
- Discussions of different mathematical approaches students may take,

- Sample student work and suggested teacher feedback to help students learn,
- Observation and interview techniques,
- Questions for student reflection,
- Guidelines for student self-assessment,
- Assistance in developing portfolios.

For more information, contact the publisher:

D.C. Heath and Company
School Division
1701 Novato Boulevard, Suite 209
Novato, CA 94947-3028
(800) 235-3565
http://raytheon.com/pub/areas.htm/

TECHNOLOGY

> In our present rush to utilize and exploit the computer, we insist on asking it questions before it is ready to answer them. The programs that we put into the computer's memory help it to answer some questions capably. The computer is good at keeping records. When we ask questions that a recordkeeper can answer, the computer serves us well. We can give a computer "expert" knowledge of a given, sharply restricted domain. If we stay within that domain, the computer's answers are reasonably competent. . . . But the computer is always likely to make absurd mistakes that reveal it is not ready to answer our harder questions because it does not know enough. (Charles Van Doren, 1991, pp. 381–382)

In the *Curriculum Standards*, the NCTM writers emphasize that calculators should be used at all levels of mathematics teaching and learning. From Kindergarten on, calculators can generate patterns, provide confirmation of conjectures about operations and procedures, and—when the purpose of the activity is not calculation per se—calculators can free up the mind (and time) of the student to attend to other important mathematical issues. In the third through sixth grades, calculators should have fraction capability so that students can use the power of the calculator to develop fraction patterns, examine reduced fractions, and convert from improper fractions to mixed numbers and back again. Graphing calculators are a newer tool that can be used fruitfully from about the seventh grade on to deal with statistical, algebraic, and geometric topics. We encourage you to attend workshops that focus on the use of calculators appropriate for your grade level and build their use into your teaching.

Computing technologies are more problematic because they are more expensive than calculators. The typical setup in a school is to create a computer lab with about 30 machines so that each student in a class can work individually or in pairs. One result of creating computer labs is that a large amount of money is spent on low-end machines—machines that really will not do what you want them to. We feel that a better outlay of district monies can be made by authentically integrating computers into the regular classroom. If, instead of purchasing a lab full of machines, the money were used to buy one really good, top-of-the-line computer for each mathematics classroom, along with a projection device or large-screen monitor, all children in a classroom can interact with something significant technologically *every day*. When a discussion on what would happen if a couple of outliers were removed from a statistical data set arises, the teacher can call for conjectures, and then the class can test their hypotheses right then instead of having to schedule lab time the next week.

This strategy for technology infusion is practical—the technology goes where it will do the most good, in the classroom—and it is cost effective. A few high-end machines will cost about the same, or slightly more, than a whole lab full of low-end machines.

While we cannot list all of the good software in mathematics, we do provide a description of a few exemplary programs that fundamentally change how kids think about mathematics.

Lego Logo

Lego Logo is a system that combines two powerful tools for children: Legos, the popular building blocks that snap together, and Logo, a dynamic system of computer programming for young children. In the initial versions of the software, Logo allowed children to give commands to a turtle on the computer screen. A child could type the command "FD 50," and the turtle would "walk" 50 steps on the screen. If the child then typed, "RT 90," the turtle would turn 90 degrees. By putting a series of commands together,

```
FD 50
RT 90
FD 50
RT 90
FD 50
RT 90
FD 50
```

the path made by the turtle would be a square figure with sides of length 50. When combined with the Legos and robotic apparatus, children can build a small robot and command it to move about the room. This product helps children learn plane geometry at a very young age from an orienteering perspective—planning, describing, and making trips. It is highly motivational, and it seems to tap into the ways in which children initially conceive of shape and space.

There are other features of Logo that are helpful to the teacher as well. We encourage you to try Lego Logo and other Logo options and incorporate them into your primary and middle-elementary classrooms. The telephone number for Lego Dacta in the United States is 1 (800) 527-8339.

Dynamic Geometry Software

In the past decade, a new type of software has been developed that helps students think about geometry from a dynamic perspective. *The Geometer's Sketchpad* and *Cabri Geometry II* are two fine examples of such tools. Students can construct geometric figures on a computer screen, manipulate the figures, and watch what happens. For example, to teach the relationship between the trigonometric ratios and the side lengths of a right triangle, a student can construct a right triangle, take the measures of the sides and angles, and use the built-in calculator to compute the ratios of the side lengths. When the student stretches or shrinks the triangle, the measures of the sides and angles are registered automatically and appear to be continuous. The calculated ratios are updated continuously. By fiddling with the computer image, the student will be able to see how at any given angle, a fixed set of trigonometric ratios exist. This is very powerful, especially for those students who are visual thinkers. *The Geometer's Sketchpad* is published by Key Curriculum Press:

Key Curriculum Press
P.O. Box 2304-ASBG
Berkeley, CA 94702
(800) 338-7638
Internet: http://www.keypress.com/

Cabri Geometry II is distributed by Texas Instruments; 1 (800) TI-CARES. Internet:http://www.ti.com/

The Internet

We strongly encourage those of you who are not online to obtain access to the internet in your schools. The internet contains a wealth of teaching resources, including demonstration and public domain software, lesson plans, videos, sounds, and images to use in your lessons. You also can access special projects like those presented in earlier sections to gain more information or to solicit advice.

For those of you who are more adventurous, we encourage you to learn how to create multimedia lessons on the internet. The markup language used to create these documents is surprisingly easy to master, and online help is easy to obtain. The internet will change the way you think about teaching.

PARENTS

When you think of resources, do not limit yourself to material and technological tools. The people you work with, parents, and your administration can serve as important support systems for your new approach to teaching mathematics.

We have worked for the past 4 years on creating and testing innovative curricula in the middle grades in the Mathematics in Context Project. After a particularly difficult semester dealing with the concerns of parents, Meyer and colleagues (Meyer, Delagardelle, & Middleton, in

Box 5 Strategies for Working With Parents

1. Be Proactive!
   a. Anticipate parental concerns, and try to address them from the start.
   b. Look like you know what you are doing. (Exude confidence.)
2. Listen to the parents.
   a. Remember, the vast majority of parents, even those who have objections to your project, are concerned about the learning and welfare of their children.
   b. Find out what they believe, what they want, and what they are concerned about.
3. Phrase your response in terms of parents' concerns.
   a. Speak or write in plain language. Define jargon so that parents will be able to understand it clearly.
   b. Do not promote your agenda over the parents' beliefs.
4. Be honest.
   a. You do not know everything that will happen, so do not pretend you do.
   b. If you can be faulted for an oversight or ignorance, admit it and follow up accordingly.
   c. Avoid being defensive, even when accusations are vehement, illogical, or just plain wrong.

5. Look for support from people within the community.
   a. Do not bring in outside people to direct efforts when local resources are available and willing.
6. Define accountability.
   a. Explain how the success of the project will be measured.
   b. Discuss the safeguards that are in place.
7. Use the power of the media.
   a. Encourage articles about your project full of pictures of students working hard.
8. Remember, you are teaching mathematics. Do not get distracted with peripheral benefits of your project. The bottom line is, "Will students learn better or not?"
9. Be prepared with visual aids and other indices of authority. (Make them readable, please.)
10. Do some mathematics with parents.
    a. Choose activities that are difficult and that tie to practical applications.
    b. Have a list of pet activities you can do on a whim to illustrate important points you want to make.
    c. Make sure some of your activities are not *number* activities.
11. Treat parents as equal partners in the process of education.
    a. Suggest ways they can help their child and their child's teacher.
    b. Provide a telephone number for parents to call if they have questions or concerns.
    c. Invite them to visit classrooms and engage in the mathematics taking place.
12. Develop "Parental Involvement Packages."
    a. These should be concise newsletters providing descriptions of the mathematics students will engage in over the next few weeks.
    b. They should include key activities that parents should do with their child at home. Encourage teachers to assign these activities as *graded* homework.
    c. They should include general news items.
    d. They should go home at *least* once a month.
13. Develop Family Math Nights
    a. Organize these events at the beginning of the year!
    a. Schedule them with regular school open-house nights or parent conferences.
    b. Structure the events to be *active* doing, not *passive* listening.
    c. Create a variety of work stations rather than working from the whole auditorium.
    d. Videotape footage of students in classrooms actively working, and show it at key points in the event.

press) decided to summarize a set of recommendations for interacting with parents that stem from the larger body of research but that focus explicitly on how to involve parents more productively in the process of change as advocated by the *Standards* (see *Educational Leadership* 51, no. 4, for a series of articles that describe the difficulties, benefits of and strategies for working with parents). The list in Box 5 may help you avoid some of the pitfalls we experienced out of our own ignorance.

FAMILY MATH NIGHTS

We have witnessed a number of schools beginning to showcase their new programs in the format of Family Math Nights. In nearly every instance, these gatherings have been overwhelmingly successful. If they are organized well, they generate enthusiasm at the beginning of the year, address parental concerns before they reach a critical mass, and inform parents about the changing nature of school mathematics and what the school is doing to be at the forefront of mathematics education.

These nights should be planned and presented by the math faculty at your school. You can contact local experts in mathematics education from a university or local education agency to steer you in the right direction, but the real communication of *your* vision should come from *you*. Many different activities should be planned, and at the end of the evening, bring closure by outlining the *Standards* and sharing the philosophy behind the activities. Remember to offer resources parents can use at home if they are interested and do extend an open invitation to parents to come into your classrooms to help educate their children.

Family Math Nights have given us an opportunity to expose parents to activities that demonstrate the concept of the NCTM *Standards*. When the parents first enter the classroom and find that we are not going to just talk about ourselves and the class agenda but that we are actually going to put them to work, they begin to squirm and sweat (except that one person who was a math whiz!). However, the activity we choose is one that is fun, done in a small group, and one that the parents can feel successful about when they leave. More important, we discuss which standards were addressed in the activity and the value of group interaction. We need to remember that parents who come to open houses are usually key people within the parent network. If we can help educate them about the value of this type of method, we are well on our way to implementing the *Standards*.

TEACHER TEAMS

Just as parents need to be introduced to your new teaching methods and philosophy, you will need to initiate fellow staff members. The staff members could be other math teachers, so your development can be done at a department meeting, or they could include the entire staff at a faculty meeting. Like Family Math Nights, the approach should be one of excitement on your part, where you share some new, fresh, be-a-kid-again learning opportunities. Most of us who are teachers became teachers in part because we love to learn, not just memorize, so let us involve our colleagues in learning mathematics again.

In your school, if you can develop a partnership with another teacher who also is willing to try a few new things, the partnership that results can be a delightful support for both of you. It is so much easier to brainstorm, develop ideas, and problem solve with more resources than you can bring to mind by yourself, and that is exactly what we are trying to encourage our students to do. It is absolutely necessary that the individuals involved in such a partnership be there because they want to, not because someone dictates that they must. When two or more teachers are helping each other achieve a goal, successes will follow naturally—and you will find that success breeds success. Others will join this support team naturally. The key is finding that one other teacher.

The ideal situation for developing collegial support would be to find another math teacher at your same grade level. If that is not possible, a teacher in another grade level at your school will work also. To make this as effective as possible, specific meeting times should be set up: These meetings should be firm commitments. This should be a priority to both of you.

Here is an example of how such a support system could work.

- Decide on a unit that both of you could do at the same time or within a few days of each other.

- Set up a daily meeting time (We meet at least 4 days per week), before school, during prep hour, after school, and so on.

- Work through the lessons that you will present to your students. Troubleshoot, add to, delete some activities, and make the unit fit the needs of your students.

- If one teacher is a bit more confident, let him or her do the lesson with the students first and test the waters for the other teacher. This way the first teacher can pass on advice about what went well or what might need to be changed to the other teacher.

- Once the unit has been started with the students, discuss what would need to be changed the next time you teach it. (Be sure to record your decisions.) Discuss what went right and what went wrong; be sure to talk about possible assessment ideas, then work through the next part of the unit.

- At the end of the unit, have a list of the *Collective Standards* for your grade level and decide which of the standards was addressed in your teaching. Brainstorm about what you feel your students have learned by that unit. You might ask them to tell you what they feel they have learned.

It is important that you and your partner share the failures as well as the successes. Most of us learn best by our mistakes, and we can help others avoid those pitfalls by discussing why something went wrong and how. As others attempt these methods in their own classrooms, be sure to offer encouragement and support.

As a second level of support, university faculty can offer assistance over the telephone, and often they are willing to give assistance as needed in your classroom. In our immediate area, we are fortunate to have the University of Wisconsin–Madison only a few miles away. Many times I have telephoned to clarify a question, and for one 4-week unit, a faculty member came out during my prep hour three times a week.

A telephone call often can help solve the problem or point you in the right direction if you are not close enough or fortunate enough to have this kind of resource. After working on a curriculum development project in conjunction with a university, I have learned that we do not use our university faculty as a resource in the ways we should. By staying in contact with them, it keeps communication open and bridges the gap between actual happenings (schools) and created happenings (colleges and universities). We both gain in the long run.

ADMINISTRATION AND THE SCHOOL BOARD

Presenting the changes you are making in a positive light to your administration, the school board really can ease the way for reform in your district. We cannot stress enough how important it is to be *proactive* instead of *reactive*. When you are challenged by people with differing views on mathematics education, work toward being positive rather than defensive. A few proactive ideas follow that you might initiate or suggest for your school.

After several other teachers have caught your excitement (in our case, it took 2 full years of work), give a presentation to the school board. When we presented the new direction we were taking to the school board, we used an activity to involve the members, much like we did with the Family Mathematics Nights. The first reaction of the members was quite similar to the parents at the open house—they were nervous!—but when they realized it was a group project and no one was going to be singled out, they all got into it. People attending the meeting were invited to join into the activities—which they did—and in groups of four or so, they tried to solve the following problem:

> If your faucet leaks 1 drop every 5 seconds, how much water will you lose in a year?

Paper, pencil, calculators, math books, dictionaries, water droppers, and measuring spoons were made available. We allowed them plenty of time to solve the problem. After discussing how and why they came to their conclusions, we reviewed which *Standards* had been touched on by this activity, and we also discussed the use of calculators. We then outlined how our mathematics program was changing, using the problem to highlight important features of our new methods. As a result of the meeting, the majority of the school board members and parents in the audience supported our efforts.

THE NATIONAL COUNCIL OF TEACHERS OF MATHEMATICS

If you are not currently a member of the NCTM, you should be. The NCTM is the largest organization of mathematics teachers in the country, providing conferences on mathematics education at the state and national level, publications on mathematics teaching and research, and policy documents aimed at reforming mathematics education. The following list of resources is by no means exhaustive, but it will get you started.

Addenda Series

The *Addenda Series* to the *Curriculum and Evaluation Standards* provides a wealth of activities for teachers to try in their classrooms that embody many of the tenets of the *Standards*. We encourage you to purchase a set and use the activities in your own teaching.

Grades K–6

For elementary teachers, there are content-specific booklets across grade levels, as well as comprehensive booklets available for your specific grade level.

Kindergarten Book: Addenda Series, Grades K–6

First-Grade Book: Addenda Series, Grades K–6

Second-Grade Book: Addenda Series, Grades K–6

Third-Grade Book: Addenda Series, Grades K–6

Fourth-Grade Book: Addenda Series, Grades K–6

Fifth-Grade Book: Addenda Series, Grades K–6

Sixth-Grade Book: Addenda Series, Grades K–6

Geometry and Spatial Sense: Addenda Series Grades K–6

Making Sense of Data: Addenda Series, Grades K–6

Patterns: Addenda Series, Grades K–6

Number Sense and Operations: Addenda Series, Grades K–6

Grades 5–8

Dealing With Data and Chance: Addenda Series, Grades 5–8

Developing Number Sense in the Middle Grades: Addenda Series, Grades 5–8

Geometry in the Middle Grades: Addenda Series, Grades 5–8

Measurement in the Middle Grades: Addenda Series, Grades 5–8

Patterns and Functions: Addenda Series, Grades 5–8

Understanding Rational Numbers and Proportions: Addenda Series, Grades 5–8

Grades 9–12

Algebra in a Technological World: Addenda Series, Grades 9–12
Connecting Mathematics: Addenda Series, Grades 9–12
A Core Curriculum—Making Mathematics Count for Everyone: Addenda Series, Grades 9–12
Data Analysis and Statistics Across the Curriculum: Addenda Series, Grades 9–12
Geometry From Multiple Perspectives: Addenda Series, Grades 9–12

Yearbooks

The NCTM publishes a book each year with stimulating chapters on mathematics teaching and learning focusing around a topic of interest. To gain more knowledge about a particular topic, we suggest you start here.

Professional Development for Teachers of Mathematics (1994)
Assessment in the Mathematics Classroom (1993)
Calculators in Mathematics Education (1992)
Discrete Mathematics Across the Curriculum (1991)
Teaching and Learning Mathematics in the 1990s (1990)
New Directions for Elementary School Mathematics (1989)

Journals

Periodicals produced by the NCTM focus on teaching and learning mathematics; and providing activities, case studies, vignettes, and descriptions of innovative programs on a regular basis. Becoming a member of NCTM entitles you to one of the following journals. Additional journals also can be ordered.

> *Mathematics Teacher* (for teachers of Grades 8–14)
> *Mathematics Teaching in the Middle School* (for teachers of Grades 5–9)
> *Teaching Children Mathematics* (for teachers of Grades pre-K–6)
> *Journal for Research in Mathematics Education* (for anyone interested in going into depth in learning the latest research findings in mathematics education; this is a technical resource)

In addition, the NCTM sells videotapes, classroom aids, and many other products that will help you achieve your goals of teaching to the *Standards*. Contact NCTM at

> National Council of Teachers of Mathematics
> P.O. Box 25405
> Richmond, VA 23286-8161
> (703) 620-9840

Keep in mind, resources are available to you. Don't go it alone without support. Find the materials, tools, and technologies you will need. Get parents, administration, and especially other teachers on your side. Join the NCTM. All of these strategies will help you teach better and enjoy it more. Ultimately, your students will benefit from your increased understanding.

Reflections

. . . if we do discover a complete theory, it should in time be understandable in broad principle by everyone, not just a few scientists. Then we shall all, philosophers, scientists, and just ordinary people, be able to take part in the discussion of the question of why it is that we and the universe exist.

Stephen Hawking (1988, p. 175), Lucasian Professor of Mathematics, Cambridge University

A TRUE STORY: RACHEL PART II

Look back to the story of Rachel in the Introduction and evaluate your answer to the self-directed question. Now how would you help Rachel if you were her teacher?

Here is what we did.

When Rachel indicated that she could not do mathematics, Jim countered with, "Ah, but that was something completely different. You have never tried anything like this. I think you might be surprised. What if you and I formed a group and work through it together?"

Rachel and Jim worked together as a team to solve the problem. He functioned as a foil for her ideas. When she tried things, he asked her why she tried it and what would happen if he changed the numbers, "Would it still work?" Together they worked through the problem using Rachel's strategies. When time came for the class to come back together to share strategies, Jim winked at Polly to call on Rachel. She was timid, but she described her thinking, with Polly asking probing questions when she was stuck. It turned out that Rachel's visual strategies were completely different from the strategies other groups came up with. Several students admitted, "Oh, now I *see* it!" when she drew a picture on the board. The class then discussed how the different strategies all pointed to the same conclusions.

It would be false to tell you that from that point on Rachel loved mathematics, that her abusive situation changed, and that her self-image reversed. It did not. However, she *began* to interact with other students more, using her own sense-making strategies and connecting them with the strategies of other students. Her performance improved, and she indicated that math was not so bad after all. For at least 45 minutes a day, she was valued for whom she was and what she could do.

SELF-DIRECTED QUESTION

1. Why did this strategy work? What strategies from this book did we use? Is there something you would do differently? If so, how?

CONCLUSION

In this book, we have attempted to provide classroom teachers with ideas that will spark their initiative to begin implementing the NCTM *Standards* in their teaching practices. Throughout this book, we have paid special attention to connecting the areas of curriculum, teaching, and assessment; if you found the distinction between the three to be a bit fuzzy, we see that as a good sign. The difficulty with writing such a book hinges in part on treating integrated features of a good mathematics program as separate topics. We hope that you have come to realize that change in any one feature of your practice will necessitate change in the other features.

Another difficulty is the breadth of coverage of each of the *Standards* documents. Some of you may be frustrated that we did not go into detail for any one content strand in mathematics. There are two reasons for this. The first is purely practical. The *Curriculum Standards* has delineated content standards for Grades K–4, 5–8, and 9–12. To treat the content effectively to the extent necessary would go beyond the scope of the format of this series. The second reason is that the NCTM has done a good job of expanding on the *Standards* documents, complete with vignettes, problems, questions, and answers specific to grade levels in their *Addenda Series*. If you have specific questions about mathematical topics, we suggest you purchase these booklets and let them be your guide.

To summarize this book into its major points, we find that the points fall into six major recommendations.

- First, a shift in practice necessitates a shift in perspective. Our beliefs about the nature of mathematics and the nature of mathematics teaching and learning must change to one of mathematics as a sense-making enterprise. Please note that this does *not* mean that we should *make* mathematics make sense for our children. Rather, it means that we should *present* mathematics that our children can make sense of. Children must be the authors of their own knowledge.

- Second, coherence in a mathematics program, from curriculum, to teaching strategies, to assessment practices, is critical for effective change to take place. Each facet of a mathematics program must feed into the others in the service of developing mathematical understanding. Replacement units, fun activities, and cooperative grouping are all good places to start, but they do not constitute a mathematics program. Groups of teachers must meet together to design and implement their vision for mathematics instruction for each grade level and across the lifespan, paying special attention to how concepts exposed to children in the early grades are developed in the latter grades and lead to higher understanding in the higher grades.

- Third, mathematics is a contact sport. Difficult concepts, multiple strategies, and especially mathematical communication are impossible to achieve if students are not required to work together, discuss, and even argue in the process of making sense of mathematical problems. This does not eliminate the need for individual work or whole-group instruction. A balance of grouping practices should be selected that are appropriate for the task at hand. This also holds true for teachers. Difficult pedagogical problems can be solved best when teachers meet to plan, troubleshoot, and reflect.

- Fourth, assessment is more important than evaluation. Develop your ability to find out what your students know. This includes their misconceptions as well as their reliable knowledge. Grading should flow from assessment, not the reverse.

- Fifth, there *are* resources available to help you teach more effectively. Tap into innovative curricula, technology, and human resources. No one can do it alone.

- Sixth, all students are capable of mathematical sense-making. Develop methods for allowing your students to use their preferred modes of thinking to add to the mathematical discussion and help them develop flexibility in the strategies they choose to solve problems.

These points were integrated into each of the goals of this book relating to philosophy, curricula, teaching, assessment, and resources. If you work diligently and intelligently on reforming your own practice, you will find the rewards far outweigh the costs. Your children will learn more mathematics, enjoy it more, and be better able to apply it to their lives. You will also reap personal benefits. You will begin to relearn mathematics from the fresh perspective of a child. You will become mathematically powerful in your own right.

glossary

Abstract— dissociated from physical reality. Having intrinsic meaning but little explicit tie to an external representation (*see* Concrete).

Activity— the mental and oftentimes physical engagement employed by individuals in the process of problem solving.

Assessment— a process of determining what an individual knows about a given topic.

Concrete— having an explicit tie to a physical representation or model.

Constructivism— a philosophical orientation to learning that assumes that individuals actively build knowledge (*see* Social Constructivism).

Content— the historical and colloquial understandings held in a field of study that delineate that field from other fields.

Context— a real or hypothetical setting within which content is embedded.

Discourse— verbal and nonverbal communication of ideas among individuals geared toward understanding a common problem.

Equity— the provision of equal opportunity, appropriate experiences, and nondiscriminatory behaviors to all people.

Evaluation— a process of assigning the worth or value of a statement, strategy, or product.

Manipulative— a concrete object or set of objects that a student can manipulate or put together to solve problems.

Pedagogy— teaching.

Phenomenological— having to do with experience. Understanding with the senses.

Problem Solving— a process by which an individual builds knowledge in a situation for which there is no ready solution strategy available.

Quantitative— having to do with numbers and number relationships.

Reasoning— a logical process of making conjectures and hypotheses and testing and verifying conjectures through problem solving.

Replacement Unit— a set of mathematical experiences used to supplement or take the place of portions of traditional curricula but that is not a coherent curriculum in and of itself.

Representation— a model, strategy, verbal explanation, drawing, and so forth, that describes the underlying features of a mathematical idea.

Social Constructivism— a philosophical orientation to learning that assumes that individuals build knowledge in conjunction with others, developing common meanings, using diverse perspectives and understandings.

Spatial— having to do with patterns, relationships, and representations of space. May include investigations of shape, position, movement, and deformation.

Standard— a statement of what is valued that is held as a goal.

Symbol— a written, verbal, or concrete instance that stands for an abstract idea.

references

Anderson, J. R. (1990). *Cognitive psychology and its implications* (3rd ed.). New York: W. H. Freeman.

Ausubel, D. P. (1977). The facilitation of meaningful verbal learning in the classroom. *Educational Psychologist, 12*, 162–178.

Ball, D. L. (1993). Halves, pieces, and twoths: Constructing and using representational contexts in teaching fractions. In T. P. Carpenter, E. Fennema, & T. A. Romberg (Eds.), *Rational numbers: An integration of research* (pp. 157–195). Hillsdale, NJ: Lawrence Erlbaum.

Brown, J. S., Collins, A., & Duguid, P. (1989). Situated cognition and the culture of learning. *Educational Researcher, 18*(1), 32–42.

Carpenter, T. P., Ansell, E., Franke, M. L., Fennema, E., & Weisbeck, L. (1993). Models of problem solving: A study of kindergarten children's problem-solving processes. *Journal for Research in Mathematics Education*, *24*(5), 428–441.

Challis, T. (1994). Assessment from a teacher's perspective. Unpublished manuscript, Arizona State University, Tempe.

Clark, C. M., & Peterson, P. L. (1986). Teachers' thought processes. In M. C. Wittrock (Ed.), *Handbook of research on teaching* (pp. 255–296). New York: Macmillan.

Collis, K. F., Romberg, T. A., & Jurdak, M. E. (1986). A technique for assessing mathematical problem-solving ability. *Journal for Research in Mathematics Education*, *17*(3), 206–221.

Conference Board of the Mathematical Sciences. (1983a). *The mathematical sciences curriculum K–12: What is still fundamental and what is not*. Report to the NSB Commission on Precollege Education in Mathematics, Science, and Technology. Washington, DC: Author.

———. (1983b). *New goals for mathematical sciences education*. Report of a conference sponsored by CBMS, Washington, DC: Author.

Council of Chief State School Officers. (1995). *Model standards in mathematics for beginning teacher licensing and development: A source for state dialogue*. Washington, DC: Author.

de Lange, J. (1995). Assessment, no change without problems. In T. A. Romberg (Ed.), *Reform in school mathematics and authentic assessment* (pp. 87–172). Albany, NY: SUNY Press.

de Lange, J., de Jong, J. A., & Spence, M. S. (in press). Ways to go. In T. A. Romberg (Ed.), *Mathematics in con-*

text: A connected curriculum for grades 5–8. Chicago, IL: Encyclopaedia Britannica Educational Corporation.

Educational Testing Service. (1995). *Packets project*. Lexington, MA: DC Heath and Company.

Fennema, E., & Carpenter, T. (1994). *Cognitively guided instruction: A program implementation guide*. Madison: Wisconsin Center for Education Research.

Freudenthal, H. (1983). *Didactical phenomonology of mathematical structures*. Dordrecht, The Netherlands: Reidel.

Gardner, H. (1983). *Frames of mind. The theory of multiple intelligences*. New York: Basic Books.

Good, T. L., Mulryan, C., & McCaslin, M. (1992). Grouping for instruction in mathematics: A call for programmatic research on small-group processes. In D. A. Grouws (Ed.), *Handbook of Research on Mathematics Teaching and Learning* (pp. 165–196). New York: Macmillan.

Gravemeijer, K. (1994). Educational development and developmental research in mathematics education. *Journal for Research in Mathematics Education*, 25(5), 443–471.

Hawking, S. W. (1988). *A brief history of time: From the big bang to black holes*. New York: Bantam Books.

Hawkins, D. (1972). Nature, man, and mathematics. In D. Hawkins (Ed.), *The informed vision: Essays on learning and human nature* (pp. 109–131). New York: Agathon.

Jensen, R. J., & Williams, B. S. (1993). Technology: Implications for middle grades mathematics. In D. T. Owens (Ed.), *Research ideas for the classroom* (pp. 225–243). New York: Macmillan.

Lesh, R. (1979). Mathematical learning disabilities: Considerations for identification, diagnosis, and remediation.

In R. Lesh, D. Mierkiewicz, and M. G. Kantowski (Eds.), *Applied mathematical problem solving* (pp. 111–180). Columbus, OH: Educational Resource Information Center/Science, Mathematics, and Educational Information Analysis Center.

Mandler, J. M. (1984). *Stories, scripts, and scenes: Aspects of schema theory*. Hillsdale, NJ: Lawrence Erlbaum.

McCombs, B. L., & Pope, J. P. (1994). *Motivating hard to reach students*. Washington, DC: American Psychological Association.

McClelland, J. L., & Rumelhart, D. E. (Eds.). (1986). *Parallel distributed processing: Explorations in the microstructure of cognition* (Vol. 2). Cambridge, MA: MIT Press.

McKnight, C. C., Crosswhite, F. J., Dossey, J. A., Kifer, E., Swafford, J. O., Travers, K. J., & Cooney, T. J. (1987). *The underachieving curriculum: Assessing U.S. school mathematics from an international perspective*. Champaign, IL: Stipes.

Meyer, M. R., Delagardelle, M., & Middleton, J. A. (1996). Addressing parents' concerns over reform. *Educational Leadership*, *53*(7), 54–57.

Middleton, J. A. (1995). A study of intrinsic motivation in the mathematics classroom: A personal constructs approach. *Journal for Research in Mathematics Education*, *26*(3), 254–279.

Middleton, J. A., Littlefield, J, & Lehrer, R. (1992). Gifted students' conceptions of academic fun: An examination of a critical construct for gifted education. *The Gifted Child Quarterly*, *36*(1), 38–44.

Middleton, J. A., & van den Heuvel-Panhuizen, M. (1995). The ratio table: Helping students understand rational number. *Mathematics Teaching in the Middle Grades*, *1*(4), 282–288.

National Commission on Excellence in Education. (1983). *A nation at risk: The imperative for educational reform*. Washington, DC: U.S. Government Printing Office.

National Council of Teachers of Mathematics (NCTM). (1970). *A history of mathematics education in the United States and Canada*. Washington, DC: Author.

———. (1980). *An agenda for action: Recommendations for school mathematics of the 1980's*. Reston, VA: Author.

———. (1989). *Curriculum and evaluation standards for school mathematics*. Reston, VA: Author.

———. (1991). *Professional standards for teaching mathematics*. Reston, VA: Author.

———. (1995). *Assessment standards for school mathematics*. Reston, VA: Author.

Pea, R. D. (1987). Cognitive technologies for mathematics education. In A. H. Schoenfeld (Ed.), *Cognitive science and mathematics education* (pp. 89–122). Hillsdale, NJ: Lawrence Erlbaum.

Post, T. R., Cramer, K. A., Behr, M., Lesh, R., & Harel, G. (1993). Curriculum implications of research on the learning, teaching, and assessing of rational number concepts. In T. P. Carpenter, E. Fennema, & T. A. Romberg (Eds.), *Rational numbers: An integration of research* (pp. 327–362). Hillsdale, NJ.: Lawrence Erlbaum.

Romberg, T. A. (1984). *School mathematics: Options for the 1990's. Chairman's report of a conference*. Washington, DC: U.S. Government Printing Office.

Romberg, T. A., & Carpenter, T. D. (1986). Research on teaching and learning mathematics: Two disciplines of scientific inquiry. In M. C. Wittrock (Ed.), *Handbook of research on teaching* (3rd Ed., pp. 850–873). New York: Macmillan.

Romberg, T. A., & Tufte, F. W. (1987). Mathematics curriculum engineering: Some suggestions from cognitive science. In T. A. Romberg & D. M. Stewart (Eds.), *The monitoring of school mathematics: Background papers. Volume 2: Implications from psychology; outcomes of instruction*. Madison: Wisconsin Center for Education Research.

Roodhardt, A., Middleton, J. A., & Burrill, G. (in press). Decision making. In T. A. Romberg (Ed.), *Mathematics in context: A connected curriculum for grades 5–8*. Chicago, IL: Encyclopaedia Britannica Educational Corporation.

Schank, R. C. (1990). *Tell me a story: A new look at real and artificial memory*. New York: MacMillan.

Siegler, R. S. (1986). *Children's Thinking*. Englewood Cliffs, NJ: Prentice Hall.

Smith, H. (1995). *Rethinking America*. Address to the National Press Club Luncheon. (World wide web: http://town.hall.org/archives/radio/IMS/Club/950712_club_00_ith.html)

Steffe, L. P. (1988). Children's construction of number sequences and multiplying schemes. In J. Hiebert and M. Behr (Eds.), *Number concepts and operations in the middle grades*. Reston VA: National Council of Teachers of Mathematics.

Streefland, L. (1986). Rational analysis of realistic mathematics education as a theoretical source for psychology. Fractions as a paradigm. *European Journal of Psychology of Education*, *1*(2), 67–82.

Treffers, A. (1991). Didactical background of a mathematics program for primary education. In L. Streefland (Ed.), *Realistic mathematics education in primary school: On the occasion of the opening of the Freudenthal Institute* (pp. 21–56).

van den Heuvel-Panhuizen, M., Middleton, J. A., & Streefland, L. (in press). Fifth graders' own productions:

Easy and difficult problems on percentage. *For the Learning of Mathematics.*

Van Doren, C. (1991). *A history of knowledge: The pivotal events, people, and achievements of world history.* New York: Ballantine Books.

Van Hiele, P. M. (1985). *Structure and insight: A theory of mathematics education.* Orlando, FL: Academic Press.

von Glasersfeld, E. (1985). Reconstructing knowledge. *Archives de Psychology,* 53, 91–101.

Webb, N. L. (1992). Assessment of students' knowledge of mathematics: Steps towards a theory. In D. L. Grouws (Ed.), *Handbook of research on mathematics teaching and learning* (pp. 661–683). New York: MacMillan.

Wertch, J. V. (1985). *Vygotsky and the social formation of mind.* Cambridge, MA: Harvard University Press.

ABOUT THE AUTHORS

James A. Middleton is an Assistant Professor at Arizona State University. He received his PhD in Educational Psychology at the University of Wisconsin–Madison, where he worked as a researcher in the National Center for Research in Mathematical Sciences Education. His interests focus on motivational processes, children's mathematical thinking, and technological innovation in mathematics instruction and assessment. He currently teaches mathematics methods for elementary and middle school teachers and graduate courses in children's mathematical thinking. In each of these courses, he strives to infuse the latest knowledge of how children learn mathematics, the tools that facilitate student learning, and ways in which teachers can design meaningful activity in their classes.

Polly Goepfert is a seventh grade mathematics teacher at Stoughton Middle School in Stoughton, Wisconsin. She has been teaching at the middle school level since 1988, when she received her second degree from the University of Wisconsin–Whitewater. Polly has presented numerous times for the Wisconsin Math Council and for IBM with regards to applying the NCTM Standards in the classroom. She also participated in a teleworkshop on the Standards through the Wisconsin Educational Communications Board, and worked with the University of Wisconsin–Madison for 4 years, piloting the Mathematics in Context curriculum. This year, she was privileged to be nominated for the second time for *Who's Who Among American Teachers.* She currently resides in Stoughton, Wisconsin, with her husband and four children